THE FLYING BUSHMAN

THE FLYING BUSHMAN

Stories from the Heart

First published in Australia in 2022

Book design by Adam Hay Studio

To purchase book go to
www.theflyingbushman.com

ISBN: 978-0-6456697-0-1

Contents

GLOSSARY OF TERMS USED IN THIS BOOK

adze: a tool like an axe, with an arched blade at right angles to the handle, used for cutting or shaping large pieces of wood; especially used for building stock yards.

agistment: when you pay a property owner to graze your livestock on his /her property

bull buggy or bull cart: usually unlicensed 4WDs that can crash through the bush and physically shoulder escaping cattle back into the mob. When they refuse to be returned to the mob they are rolled over and tied up and picked up later in a 4WD crated truck often called a bull cart.

bull hitch or cob and co hitch: when a single desired length of wire is doubled onto itself in half and wrapped around whatever you wish to twitch together. A fencing pliers handle or pipe is placed through the eye and pulled tight against the other end of the wire, twisting tight around it.

cleanskin: unmarked animal in terms of registered earmark or ownership.

coachers: mob of quiet, relaxed cattle usually on the tail of the mob, less afraid of and used to stockmen and being handled. These are often used for hiding up wild feral cattle for protection until they settle down.

crib: lunch bag or tucker.

driving or droving: driving is moving a mob of cattle into the yards in ten minutes whereas droving is days, weeks or months, often between different geographical locations.

dunnie: an old toilet out the back of a homestead.

having eye: a genetic trait bred into working dogs connected to their complete and absolute concentration on the animal they're working on.

goat panels: metal panels approx. 1.2 metres high and 1.8 metres long with horizontal parallel bars fitted close together to avoid escape, in turn attached end to end with strong connection points.

hot and silly: when most animals, but particularly wild cattle, run too far and have a raise in their body temperature, they can become more difficult to handle and to settle down. That's why you try to avoid running them too long before enforcing a walk for them to cool down.

micky bull: a young, unmarked male entire.

old piker bullocks: old, castrated male cattle, possibly over five years old that may well have escaped previous musters and usually weigh over 600 or 700 kilos live weight depending on seasonal conditions.

panican: metal bush cup.

RMs: R M Williams made riding boots, called RMs for short.

road trains: large stock trailers to transport penned livestock long distances, often hooked together and pulled by a single prime mover.

shouldered back into the mob: this means the horse physically uses its chest and or shoulder against the shoulder of the beast to force it to head back towards the mob.

smoko: an Aussie bush term for morning/afternoon tea break or to stop work for a cuppa.

spring-fed little pool: where the water seeps up from under the ground for livestock to drink and survive on; some of these don't last all year.

taking cattle forward and keep taking cuts: taking bites or groups of cattle forward into a yard, so the others are inclined to follow.

yaki: a cry given by a stockman to move or scare up a mob of livestock into a yard.

CHAPTER 1

The Goats that Got Away from Shaughnessy's Pool

Ron Rogers from Cary Downs Station southeast of Carnarvon in Western Australia (WA), was a hard-working bush bloke who had been given little and didn't expect something for nothing. He had been a shearer, and, along with his wife Margaret, raised their children on a tough block of dirt that didn't give you much in return, but you had to admire their work ethic. It's just that some properties are better than others, with better grazing rangeland type and higher livestock carrying capacities, leading to higher incomes. Others aren't so lucky.

When he rang me one night and said, 'Greg, I was wondering if you would like to have a go at trapping those goats over at Shaughnessy's Pool just off the Wooramel River', I became suspicious. Not in the wrong way; Ron was a good man. It was just that I knew he wasn't able to give much away and figured there must have been a catch somewhere. I wouldn't have expected him to give a few grand of goats away if he could have got them himself.

He said, 'I put a bloody trap up out there, and the bastards keep on getting out on me and I can't quite work out how.'

I thought this might well be an exciting bush challenge requiring lots of sweat and effort. I didn't give the challenge enough credit.

Now I knew Shaughnessy's to be a rock pool in an isolated rough, rocky area on the north side of the Wooramel River and, although I hadn't been there, I knew roughly where it was and had heard about it from old Frank Shaughnessy himself years beforehand. Frank was an old Aboriginal stockman who was on the cover of my first book with me way back in 2016. He was a marvellous old Yamatji man, and I'm not sure how the pool managed to be named after him, but I was aware of it and where it was.

I thought I might first take a run-up on the motorbike and look at the situation for myself, before making any commitments.

After we completed some jobs on Ballythunna Station, my adjoining property, I said to Steve, my Kiwi jackeroo, let's go and look at this Shaughnessy's Pool that Ron is talking about and see why his trap hasn't worked for the intended goat collection.

It was some 20 kilometres away from where we were, so we made tracks, and when we got close, I said to Steve, let's shut the bikes off and walk in without disturbing anything to ascertain the situation. We walked in the last kilometre or so, and the first thing I noted was some billy goats coming out from the pool with bellies full of water, indicating they were getting out somewhere and escaping after having a drink.

We tried to avoid these escaping goats seeing us in order not to spook them and quietly made our way towards the pool where we could poke our heads over the hill without being seen to survey the situation.

A fault develops in the rock in these type of pools in the centre of a rocky hill or breakaway where the water has broken through the surface of the hill and eroded away the rock over thousands of years. Here, the waterfall drops into what could be referred to as a natural amphitheatre below, some 30 or 40 metres in diameter. A waterfall used to run into this eroded area, surrounded by cliffs some 6 or 8 metres high at its highest point, and down onto a section of sand area below. From here it would have overflowed into a creek system which eventually ran into the Wooramel River. Where this water from the waterfall hit the sand at the bottom of its 6- or 8-metre fall was a spring-fed little pool, only a few metres wide and perhaps a metre deep, which was where all the goats were watering.

There was a secondary waterfall, which was like an overflow or bypass when there was heavy rain. It ran around to the side of the main waterfall and had another exit some 5 metres away from the first and onto the sand below. I mention this specifically because, as you'll see, it's very relevant to the story.

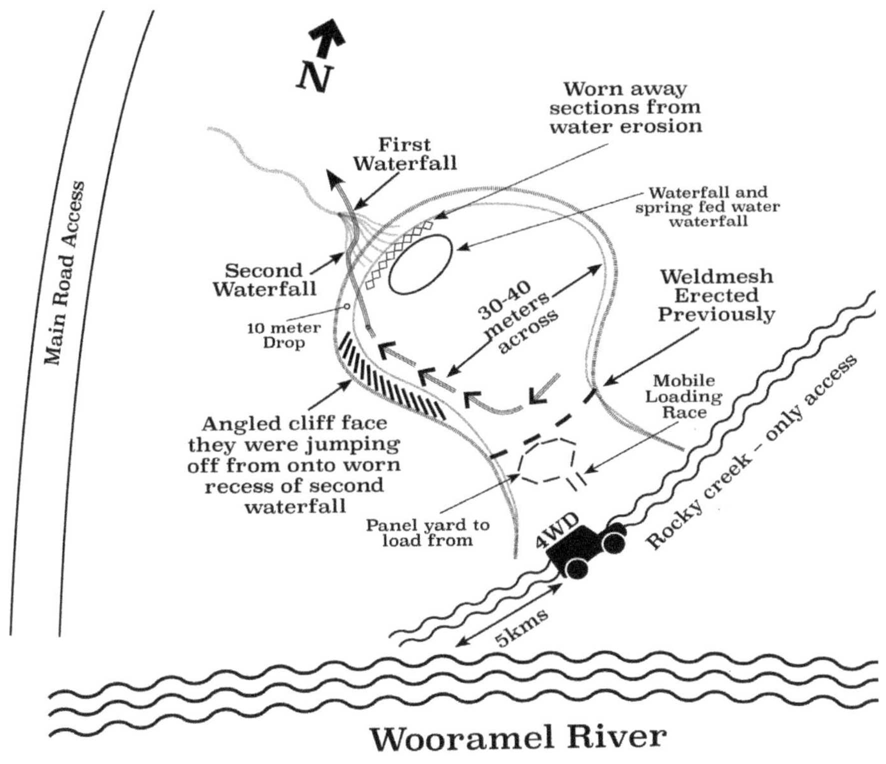

This whole area was very rocky and extremely inaccessible except by motorbike, foot or horse, which helped me understand why Ron had freely given away this opportunity.

Although an old 4WD could possibly make the journey very carefully – just.

Not only was it a matter of successfully trapping the goats, with good yards and equipment that first needed to be shipped to the site, but then the goats had to be successfully loaded from a very inaccessible and isolated place and transported many kilometres to trucking yards. Like a piece of cake – yeah right!

And the few bits of old weldmesh that Ron was presently using to restrict and trap his prey were not working very successfully, hence his frustrated phone call. To his credit, he did say I could have any goats I caught, but it was dawning on me there was a legitimate reason they were being given away.

Nevertheless, as my father always used to say, 'You can't look a gift horse in the mouth'.

Little did I know, I would have been better to dispense with the horse and walk away!

But he also used to say, 'Faint heart never wins fair lady', so I figured I better try and find this fair lady.

What I needed to see for myself was exactly how and where these goats were escaping from the trap that Ron had set up. I respected the fact he was not a fool and had plenty of experience with livestock so understood what was needed to block their escape, and on the surface of it the trap looked to be doing at least part of the job, because there were some goats still inside the trap, however we needed to find out the escape route they were using.

As we peered our heads over the top of rocks, we could see 150 goats down in the natural yard below. Over two-thirds of the yard was enclosed by the sheer cliffs of the breakaway the waterfall had eroded into, and the small section of the sandy floor running into the overflow exit to the creek Ron had closed off with weldmesh over a metre high, and some of the goats were trapped and unable to escape from there, at the minute anyway.

But close inspection was needed and given our prior knowledge that especially big billy goats were escaping from this enclosure, we watched more closely to learn more.

We could see the weldmesh trap gates that Ron had been using, which was a section of weldmesh a metre by a metre made into a V shape, so the thirsty goats could push in and open it enough to get past and not be able to come out the same way. And that seemed to be working quite effectively and the goats were not escaping back through there. And then I realised what was happening and their means of escape. And it was quite an amazing revelation to me, or to anybody for that matter.

These goats, mainly billy goats (large males) were lining up like kids in a playground, one behind the other, and taking turns getting a run up along the sand for the longest possible distance, probably 20 metres. Then, using one section of a small side of the cliff wall that wasn't completely vertical (perhaps a 40-degree angled ledge a few metres long), but still at least two metres from the top of the cliff face, they bounced from there up to the waterfall or the recess above that to their escape. This ledge was under a shoulder of the cliff face, so they could not jump from that shoulder to the cliff face directly above to escape, but only across to the worn indentation in the rock the water had made at the second waterfall.

They were using their forward momentum to get that run up because, as some proved while we were watching, if they didn't have enough speed and momentum, they couldn't make the height of the final ledge at a 45-degree jump around 3 metres above and away.

Seriously, it was nearly an unbelievable feat that they were physically achieving. Even after working with goats all my life and having great respect for their self-preservation, I was astounded. And unless I had seen it with my own eyes, I would have never believed it was even possible to achieve.

I am sure they had learnt this lesson some years previously, possibly over generations, when they had been trapped in this yard. Because, I was sure, this was a learning that had happened over a long period, certainly not in five minutes. Maybe hours and days in this yard had given them time to work on these escape routes, and once some had seen it done and achieved, others would try and try again, until successful.

And some of the goats didn't make it on earlier attempts either, with many going back and back until they were successful, and then some still didn't make it from our hour or two of watching. Talk about survival of the fittest. That was it right there.

But I certainly would have liked to see the first goat, who not only thought up the escape plan but managed to execute it successfully. It was an amazing feat to see, even after many years handling all sorts of livestock, the intelligence required was unbelievable.

So, all buoyed with excitement that I now knew their escape route, from our brilliant watching in concealment, I told Steve we would go home and later return with what was required to combat their escape, and imagined a considerable number of trapped goats in the yard, after a few days of non-escape and complete capture. On the tracks, there were many goats watering at this pool, and it was the only water for miles around.

So we returned the next day with several goat panels and other required equipment to complete the job. We installed one particular panel which blocked the exit on the second worn section of the second waterfall in the stone. I didn't have a chance to see goats attempt to get past this escape block we had activated, but it would have been impossible, even for them, to bypass it successfully. And we tidied up the weldmesh trap entrance down on the sand to make it a bit stronger and more effective, making some other reinforcements using our common sense as to what else was needed with these smart so and so's.

The pool and trap site was only accessible very slowly in a 4WD in low along the creek (as per the diagram included on P5) with a tray top without even being able to pull a trailer because until you try it in that situation, you don't know for sure. But I could drag my cattle-loading race on wheels behind the vehicle over the very rough terrain. A cattle-loading race is often used in these circumstances because it is entirely mobile, and, because it is much broader and applicable to loading cattle, it means billy goats don't get their horns entangled and hence makes easier loading.

Well, that's the theory anyway.

Because I was unable to get the trailer very close to the trap site, I was planning to ferry the goats, once loaded, a few kilometres to the trailer ten or fifteen at a time and then, once the trailer was loaded, on to trucking facilities some 30 kilometres away.

All was in readiness for the big mob to be nicely entrapped in my yard on my return in a few days. They were being trapped onto water which had been working successfully, so they wouldn't be perishing until my return. A few days later and all excited, I duly travelled back to Shaughnessy's Pool to see if we had been able to outsmart our goats.

For some reason, Steve was unavailable so there were just the three of us to do the job, me, myself and I. They're very good as they don't pinch your beer in the camp, but on the downside, they usually leave all the work to me.

I was pleased to see that, still some distance away from the trap yard, I hadn't seen any goats around full of water which was a good sign. In fact, I didn't see any goats at all, which hopefully meant most were caught in the trap and unable to escape.

To my pleasant surprise, when I stopped the 4WD some few hundred metres away from the yard and walked in, I could see 300 to 400 goats in there and very few outside.

We had found out their escape trick and blocked it successfully, it seemed, and I was delighted at what we had managed to achieve. It was not just the potential financial bonus, which was great, but we had outsmarted them at their own game with some pretty good bush ingenuity.

But I was concerned about getting the large mob into a smaller yard first from the natural amphitheatre yard, and then, with more difficulty, getting them onto the Toyota tray top at only twelve or fifteen at a time. I had my mobile cattle loading race, but I needed to be able to get it close enough to the yarded goats over and around huge boulders to be useful.

After gentle shepherding and patience, I managed to yard the whole mob off. I calculated well over 300 goats in a tight yard of goat panels that I had specially designed and built years before, which were higher, to avoid goats jumping out, as they can readily do, and with horizontal rails closer together to prevent kid goats escaping. And because this yard is not pegged to the ground and is flexible to mould into any shape, its strength comes from its end-to-end connection of panels remaining vertical, strong and upright. With this number of goats, it was probably around 10 to 12 metres in diameter and contained a large number of older billy goats that had been in these rocks for many years and had also managed to keep the dingoes at bay and survive. So, they were wild and feral without any please or thankyous; tough nuts at around 100 kilos plus each, and some with horns a metre across. These also took extra room in the small yard, their horns knocking against each other whilst jostling to get further away from me on the far side of the yard, which was already stretching and groaning under these powerful animals' considerable pressure and strength. But I was confident it would hold up as it had done before.

Because I had all the goats contained and encircled now in the large hexagonal panel yard, I needed to remove Ron's weldmesh that was restricting the goats from running towards the creek, the way the water would flow after rain. I needed to get my loading race and 4WD closer to the yarded goats for loading. This also meant I was effectively relying totally on my panel yard now to hold and contain the goats, as I had removed Ron's weldmesh as the second barrier, which I was confident about after doing it many times before in similar situations.

Especially with all that help from my friends, me, myself and I.

I removed all the mesh and steelie uprights to get my mobile loading race closer to where it was needed and again noted the portable yard flexing as the big goats were trying to move away, pushing through the mob, away from the encroaching vehicle noise. I'm sure most of them had never experienced vehicle noise in their lifetime. This was a very rough, secluded and isolated block of country with very few roads even close, so this would be a frightening and restrictive experience for them. But the yard held fast as I had hoped and expected.

But to get the yard into a funnel-type shape, ready to connect up my loading race and then load onto the crated tray top vehicle, I needed first to erect a small separate forcing yard, running from the yard the goats were now in, to receive just enough goats for the load – approximately twelve to fifteen, depending on size.

I was thinking about the thirty or thirty-five trips the three of us had to do, back up to the better road and level ground to the trailer to offload, then onto the trucking yards. It was just as well those billies were worth at least fifty bucks a head, I figured, as it seemed to make the enormous potential task a little easier to consider.

But perhaps I should not have gotten too far ahead of myself.

The holding yard buckled and groaned as it twisted its shape somewhat from the considerable pressure these extra-large billies applied from inside, pushing and forcing each other and trying to get further away from me. Some of them jumped on top of others to get further away, even though I was walking around as quietly as possible so as not to startle them more.

I realised my close pressure was affecting them but had no other option but to try and get them loaded to lighten the yard off, and it was getting hotter – above 40 degrees centigrade now, around 10 am and it was only getting worse the longer I left it. So as usual, time was of the essence.

If I could at least get one load away, that would lighten off the number in the yard and let them have time to settle down in their new environment of the panelled yard. That was the plan.

With the last few remaining panels I had, I gently started to build my very small forcing yard between their yard and the loading race, and, in so doing, had to lift some of the panels in the air to get across to the other side of the holding yard from where I had been working up to that stage.

That was enough. That's all it took to hit the fan.

Some of the goats that were closest to me rushed away from the raised panel I held above my head and jumped on top of the goats already tightly bunched on the opposite side of the yard from me.

Now, although the panels stood 1.2 metres tall, which was usually enough even to stop jumping goats, when a few of them fleeing my presence were actually on top of other tightly packed goats, it meant they were able to look over the top of the panels to freedom.

Well, you don't need to tell a goat twice.

And so a few started to jump over the yard panels and away from the backs of their mates, still squeezed up against the furthest side of the yard from me. I could only watch in that instant as quite a few had seen what was working for their mates and were keen to join in, getting a running jump off their mates and making off out into the bush.

However, before I could run around to block them and attempt to scare them back from jumping, the complete panel yard started lifting off the ground right in front of me because of the extra applied force of the goats in the yard wanting to follow their fleeing mates. More of them collectively were applying pressure to the top of the panels on the other side of the yard instead of the usual base and centre. And hence their force was lifting this large hexagon-shaped yard well into the air.

I tried, of course, to take the lesser of two evils and hold down the yard on my side to the ground, hoping the jumping and escapes would minimise and cease in time, but the reverse occurred as more goats saw more escaping and wanted to join the lead and continued their flight. By this time, the pressure and force from the fleeing mob was so great that I found myself hanging onto the base of the panels more than a metre in the air, in furious disbelief, and losing ground progressively. I knew if I let go of the panels on my side to run around to the other side of the yard, the complete yard would entirely flip over, which would not apply any restriction to them from fleeing at all.

By this stage, over half the yard had escaped, and the rest were extremely keen to follow their mates to freedom. And by now, they only had a 45-degree panel 1.2 metres high to jump over. The panels had stood firm end to end and were not broken, but it was just that that collective force from the whole mob against one side of the yard at around one metre level in height just meant the entire yard rolled over.

When those goats saw their mates disappearing over those panels, they didn't need to be asked twice. Watching their mates escape and following their example had worked well in the past for them, so why wouldn't you?

I may have said a rude word, you never know, even a few, as I was hanging like a circus clown, suspended in the air with everything bar the big clown boots on, watching all my goats run off into the distance.

In disgust, I let go of the yard panel I was holding and it proceeded to flick further into the air, allowing the side of the escaping goats to let even the last few to freedom, bar the few left that were squeezed in the crush and were getting their breath back like me, before following their fleeing mates. If someone else had been there, one of us could have run around to help stop the flow and scare the animals back, but what can you do by yourself?

I went over to my waterbag in the vehicle and, as I drank, wondered what the hell had just happened. Five minutes ago, I was looking at a win by completing a hard and challenging job to make some serious money, perhaps 4 or 5 K, remembering that the average price per head would have been closer to $20 per head, although the billies may have been $50.

As they say in the classics, you can never count your chickens until they hatch.

I was so disappointed and disillusioned after all the set-up and effort required to achieve the goal. Then to watch the goats all escape meant I had accomplished nothing. I had used all my bush skills to watch them, block their escape route in Ron's original trap and had them in the yard, with few outside. So, the hard part should have been done.

Anyway ...

After some hours at 40-plus degrees, I packed up my yards and gear and went home. I should have stopped and boiled a billy can, but I just wanted to get out of there. Those other two didn't help much either, and they're a sulky pair of b–.

This business above with me, myself and I, is entirely tongue in cheek. Right.

From around the 80s to, say, 2010, and in some cases beyond, many station pastoral properties in WA have had severe droughts or been forced out of sheep by dingoes, and hence only been able to carry limited livestock, meaning low cash flow and thus not much staff, if any. So it is pretty standard for one person to be running and responsible for half a million acres of country and its infrastructure on their own. Hence the help from me, myself and I.

Who Knew Goats Could Swim?

Kalbarri itself is a beautiful little tourist town located at the mouth of the Murchison River on the midwest coast of WA. The station property or station lease that runs into Kalbarri from the north is Murchison House Station, which used to be owned in the 70s and 80s by a man known simply as the Jah – the Niamsam of Hyderabad.

And as per the article on Google below:

Here is the sad history of Mukarram Jah, 'the Eighth Nizam of Hyderabad, Regulator of the Realm, the Victor in Battles, descendant of the Viceroys of the Deccan and heir to India's greatest dynasty since the Mughals', who went from being the richest man in the world (with swimming pools of jewels) to the owner of Murchison House Station near Geraldton and finally, as the author reveals at the beginning of the book, to a two-bedroom flat in Turkey.

I had permission from National Parks to muster goats off the land, as they were the body in charge of mustering the national park that adjoined Murchison House to the north, with the rabbit-proof fence dividing them.

One day, I was out near the Murchison House boundary and by complete fluke ran into the Jah, as he was called by the locals, who happened to be driving around the property in his brand-new Range Rover with two or three gorgeous bikini-clad ladies in tow. As a young man of twenty-something, I can tell you it was a beautiful sight to behold, way out there in the scrub, and certainly very unexpected, and I'm sure I nearly fell off my motorbike in love, maybe.

But I'm not here to tell you about that. I'm here to explain a fantastic feat that I saw goats perform, nearly as amazing as the feat they achieved in the Shaughnessy pool story above.

One afternoon, I was hanging my feet over the cliffs overlooking the ocean near the 'rabbitie', the rabbit-proof fence north of Kalbarri, and pleased to get the start of the sea breeze in my face. It wasn't long before a small mob of goats, probably fifteen or twenty, came down onto the small section of beach in front of me and, having a fair bit to do with them over the years, I guessed they looked like they were on a mission. I was some few hundred metres away, and kept still and unmoving, not wishing to attract their attention.

There was no feed down here on the beach that I was aware of, so I wondered what they were doing. I was sure they couldn't drink seawater, so the fact they were there made me think they had a goal. They kept on walking down the beach, obviously on route to somewhere and then stopped down in front of me, just out of the water.

Now I knew goats typically hate the water or getting their feet wet, as I remembered them for years in a trap yard; when you cleaned out the trough overflow, they would make sure they didn't step in the water, even when running away from you in haste. So what were they doing here?

After a minor delay, an older billy goat walked to the lead of the mob and then walked out in front of them directly into the ocean, which astounded me. I had never seen a goat in water, let alone walking out to swim in it with his legs not touching the bottom.

He continued to swim out further, some 30 metres maybe, and then started to float around in circles with his head down in the water, raising it occasionally for breath.

For me to see a goat do this was absolutely amazing.

It seemed he had found the spot where there was a freshwater-fed spring or indeed drinking water anyway because once he submerged his head to drink, the others got the message and swam out to join him, and all were swimming around to find the freshwater.

After a while and having had their fill, they swam ashore and snorted the seawater out of their nostrils and dried their wet bodies on the beach for a while before gathering up and walking back up the beach to where they had come from and disappeared. I sat there amazed. I thought I had seen most of the surprises goats had to offer, but you never stop learning in the bush.

For quite some time, I've only shared this experience with close family members and friends because I knew that many – even those with considerable experience with goats – would call it bullshit. A goat never goes near the water other than to its edge to drink if it's fresh or comparatively fresh, they would have said. But a few years later, I told the story to a truckie I was with from Cream Transport (who used to cart my livestock), having the opportunity to catch a ride to a funeral with him. He said he had seen the same thing happen up at Exmouth Gulf with a larger mob of goats than mine when he was camped on the beach fishing.

So that's what happened, and when they tell you cockroaches will be the last thing on earth after a nuclear explosion, I say rubbish.

The goats will eat the cockroaches to survive.

CHAPTER 2

Do We Really Understand the Bush?

People may think they have a concept of the real bush, but perhaps they are deluding themselves. Many things identify the bush and the bush people and even those, of course, are changing and updating all the time. Even just twenty years ago, the thought of a bushman being able to check his stock market investments on a mobile phone from the back of his horse on the tail of the mob would have been incredible, and yet it is a reality today (providing one got a signal, of course – don't get me started there!) The bush, like everywhere, has moved into the 21st century.

It's probably challenging to get into that bush state of mind when you live in a unit in central Perth or Sydney. I am always honoured when testimonials of my books say, 'You take me to that place, and I feel I can touch it.' These comments are particularly gratifying because they mean I have achieved something I always strive for when writing: I want to move and touch people. Making them laugh or cry has an impact and perhaps gives them something to ponder. If a reader does you the honour of wanting to read your book, I believe you have a responsibility to entertain them and capture their attention and imagination and to have an impact on them. Make them remember the experience they had when they were engaged in your book.

The fact is that most Australians now live in the major cities dotted around our vast coastline, and it is not their fault if they don't touch, feel, see or understand the Australia most of us originated from. It's not right or wrong – it just is. And, if they don't need to experience that lifestyle, is there a reason why they should? Well, I believe it is vitally important for people (all people) to experience the bush and the outback because, hopefully, it impacts them. The bush has a truly positive impact on human beings, I'm absolutely sure.

Marianne, a Swiss lady I know (she has been in Australia for many years), made a trip by herself in her 4WD, partly to follow some stories and places mentioned in my previous book, *The Flying Bushman: An Australian Story of Life above the Land*. She travelled out through Yalgoo and into the Mount Magnet area to a Station Stay for a few days, and then on to the Murchison Settlement and past my old home, Curbur Station. Then she travelled on to Gascoyne Junction and Carnarvon. She told me she felt like she was reborn after that experience.

The kind people she met, who were only too keen to help, the wilderness and quiet, the crackle of the mulga sticks in the campfire and their smell, the rainbird making his song before the showers came, the photos of animals she took (she's a professional photographer) – it all had a major impact on her. A giant eagle's nest only 2 metres from the ground amazed her; I don't think I've seen one that low myself. They are usually perched way up on high somewhere, where they can see predators and food from miles away.

We discussed just how clear and focused Marianne's mind was on her arrival back in Perth after this bush trip and the fresh air. She said it was like all the cobwebs had been blown out. She immediately decided to change her job and make bigger and better plans for next year. Her vision was clear.

Here's another example: I had just taken my twelve-year-old grandson Robert to Wagga Station Yalgoo, where my brother manages the station operation. One morning we went out to check for some cattle tracks, to keep the cattle from being struck by the numerous ore-hauling road trains on the highway. We were passing an extensive range of granite outcrops known as 'The Brothers' when we noticed a pair of eagles who had just made a kill of a young rock wallaby joey. The pair had collectively hounded the mother until she dropped her infant in flight, as they do, and were in the process of finishing the carcass when we disturbed them.

I thought to myself, 'What a marvellous natural experience for Robert to witness.' We saw the eagles circling from a distance and then saw the result of the ravaged carcass on our arrival. How many times would one get the opportunity to experience that? This example may be a bit gruesome, but it's life in the bush and the real world, isn't it?

There is just so much to take in with all your senses. The sounds, the smells, the different views. Just being one with nature so you can sit and take it all in, realising the world's vastness all around you and what small cogs we all are in its midst. It makes you humble and yet, at the same time, fully alive.

The recent bushfires and floods are the first time Mother Nature has directly impacted many Sydneysiders, so perhaps it's an apt time to consider what's happening in the real world. They have my complete sympathy for what has been thrown at them in recent times. Most bush people have felt nature's wrath at some stage or another.

And so it is with the bush: you can't know what it's like unless you have heard that eerie sound of a wild dingo howl late in the night from your swag under the stars, especially as a child, when it puts shudders down your spine. And you won't know the taste of cool fresh water from a mulga-shaded creek on a hot summer's day, water that holds nature's natural minerals, extracted from the oldest rocks on the planet. As well as some kangaroo manure, of course, but that's OK because it mixes up and you don't even notice it. There's just that overpowering freshness and potency from nature that bottled water will never achieve.

Have you heard the rainbird sing his melancholy song before the approaching rainstorm? It's such a powerful, intimidating call, but also beautiful when you know what he brings with him.

Have you watched, smelt, heard or felt the warmth of a campfire crackling gently at sundown, taking you into the gloom of night and confirming that 'out here' it's just you and Mother Nature, the big world out there? And it feels like a very little 'you'.

Most people in the regions regularly visit the cities for business, health and other purposes before hastily returning home.

Does it matter if regional Australia dies slowly?

Well of course it does matter, and many people are trying their hardest to keep following their dreams in rural Australia under challenging circumstances, and it is thanks in part to many of those producers and their families that Australia is where it is today.

The service industries in the big cities are indeed providing a substantial 'service', but they are not producing, growing, building or making something out of nothing. And that's why so many young people are out there trying to experience the regions and outback Australia. The backpackers come from across the world to experience isolation and to work in the outback.

The genre I write has a large target audience of readers, even if they are not there in the outback themselves. Those readers have often heard their grandfather or uncle speak across the kitchen table on their country experiences.

As shearers, fencing contractors, drillers, stockmen or horse breakers, we come across amazing stories of how things were and, in some cases, still are. That's our Australian history and where we came from, and we have every right to be incredibly proud of it.

It's not just the country that impacts you, but the people of all nationalities and creeds that have come here to stay, call the country home and carve out a crust. They have been impacted by that bush environment themselves – who and what it has encouraged them to become and the way it has influenced how they think.

Nearly all jackeroos (mainly, historically) and jillaroos (in more recent times – the last thirty or forty years) spent their younger days on stations somewhere, learning about nature and life to draw on for their future. Even station kids themselves went and jackerooed over east or somewhere.

It doesn't matter if you're a lawyer, doctor, dentist or bricklayer, bush training in your youth will always stand you in good stead to handle whatever comes along in life. It instills in you a lot of common sense and self-confidence.

I have known people from across the world, many of whom have seen all the well-known tourist destinations on the planet, who love returning to the Australian bush and outback for its magic. One way of touching that magic is talking with people close to the land who live on it daily.

A bush bloke who readily comes to mind is John Leeds, who is involved with his sons in running a successful livestock trucking business north of Perth in WA. John has been directly involved in the pastoral industry and regional WA all his life. He was born and bred to it. His family started near Cue in WA in the late 40s and been involved in the pastoral industry ever since.

I first met John and his family in the early 80s when I helicopter-mustered his property, Mini Creek, in the Gascoyne Region of WA, not far from Carnarvon where he started with his first Dodge truck to cart livestock in 1972. There is hardly a station or farming property in northern WA that John has not carted at least one truckload of cattle, sheep or goats out of or into, or dropped some freight into, over the last fifty-plus years. And just for the uninitiated, this means covering thousands of square miles (not kilometres) of country, meeting and dealing with all its characters along the way. He knows everybody and every station, from their homesteads and their cattle yards to their old dunnies out the back that he's had to use. And he was doing a job for them, often in the most difficult circumstances, including substandard yarding and loading facilities, thunderstorms, droughts, inaccessible roads, floods, bogs, fires, and, at times, the crankiest and weakest of livestock, never mind some of the people he had to deal with while trying to get the job done. He needs a medal, seriously.

But he's a bush bloke and takes it all with a grin. However, as he gets older, I'm sure he suffers fools less gladly.

In all that time and all those millions of kilometres travelled, he has always arrived on time, even in the most inaccessible of places. He also during that time in the bush developed great relationships with some great old Aboriginal stockmen and horsemen that he respected and worked with, like old Tony Moncrieff. And that mutual trust was developed over many years of working with them to know they were men of their word, and when they said there was a particular problem somewhere on the property, they earned your attention and respect.

Another man of the bush with similar qualities and bush-trucking experience is Herbie Richards from Mullewa. I have had great personal dealings with Herbie and his son Alan for many years, some of which were in challenging circumstances. Having to truck and handle weak sheep and livestock to transport them to better pastures far away during years of consecutive droughts is heart-breaking. It stretches anybody's strength of character and patience to the absolute end.

And it happens because some animals are in poor condition or haven't the strength to handle the stresses and strains of transport and loading and unloading as well as travel. Working with these animals day after day can be difficult. Then add in attempting to keep to market and travel timetables for trucks and transport, even being required to be somewhere else hundreds of kilometres away in an hour's time where staff are waiting on hand to load and the livestock you are trying to load are falling over at your feet, and you still have 500 to load. It tests the patience of the best stockman, I tell you, and sometimes the only thing you can do is pull off and have a smoke, if you do that, because honestly it can get the better of you if you don't switch off. And it's important here you don't take your frustration out on the animals, which I never saw these blokes do over many years. It's not the animal's fault that it is in the condition it's in.

In the company of these experienced, mature bush men, I drop into that well of understanding and stories that instantly take me back to those places. You can nearly see them and smell the rubber burning as they speak.

Don't you think they've sat around the odd campfire in their time? Or had the odd adventure? They have patched up trucks on the side of rough roads with whatever material they could to get their clients' livestock to market. There are stories of using a wooden guidepost on the roadside to bull hitch over a broken truck spring, to get livestock to market down the road a few hundred kilometres until proper repairs can be done. They are so present and involved when experiencing those dramas, and they inject them into your life when they tell their stories to you.

As I was writing this chapter, Herbie Richards rang me out of the blue on a completely different issue. While I had him on the phone, I mentioned that I was writing this chapter for the book and that I would be using him and John as examples of experienced bush people. Directly off the cuff, without any prompting, Herbie started to relay some memories he and I shared that I had mostly forgotten. Although they happened forty years earlier, he recalled them like it was just yesterday, in every minute detail. And this recall illustrates exactly what I am attempting to describe when it comes to some of the traits of these bush people. They are so present in what they do or have done.

He told me of a night he remembered well when he and I (but mostly he, with his mechanical expertise) had worked on my helicopter at Curbur Station, where I based my helicopter mustering operation. I had recently flown it back from Jandakot after a 100-hour service. The oil pressure light on the dash had started to come on occasionally, and I was obviously concerned about it.

I rang up the engineer, Burt Jennings at Piper West in Perth, to discuss the situation. He said they had recently replaced the oil-cooler unit and that it would be worth checking that first, under the circumstances. But he also said that any significant mechanical adjustments would need to be completed by a licenced aircraft mechanic. Burt knew me well and knew where this was going. We had done thousands of hours together, as it were. Remembering that we were 1,000 kilometres apart, I mentioned I had a good mechanic handy, and we could at least see if that was the problem. I got the usual concerned chuckle down the phone from Burt, saying he could feel another sleepless night coming on and was wondering if he would still have his Aviation Engineer's ticket the following day, as it was illegal for an unlicensed mechanic to work on a machine and if Burt could be seen as having given mechanical advice, it would look bad for him. But he's well and truly retired nowadays, and I'm sure he would have a giggle reading this. I hope the Department of Aviation doesn't see this! I don't think they can read anyway, so it's OK.

Anyway, with Herbie's extensive toolbox and his excellent small hands, which had fixed many mechanical dramas in tight places over the years, we found that the gasket had been placed upside down so that the holes in the metal for the oil flow didn't match with the gasket. Thus, the oil flow had been restricted, causing the low oil pressure light to illuminate. Once we saw the problem, we were convinced that was the issue.

After confirming the problem, we replaced the gasket in the correct position. According to Herbie, we took off for a check ride with him as a passenger at 11 that evening. I just can't believe we ever would have done such a dangerous thing, but I'm backing Herbie's memory. You see, these helicopters are not licensed for night VMC operations (flying at night), as they haven't got the required navigation equipment. It all checked out OK under load as we needed to test it, but I'm still saying Herbie's watch must have stopped.

A few days later, according to Herbie, I flew into the homestead while he was there doing some mechanical work, to say that I had mustered up a mob of feral goats on the west side of the property, some 30 kilometres out, near Wheendong cattle yards (look up Wheendong Claypan on Google maps: https://earth.google.com/web/search/wheendong+claypan+in+wa/@-26.48531526,115.71418502,280.83566344a,1438.67067642d,35y,359.99992416h,0t,0r/data=CigiJgokCfREHT6snfQ_EWKs0et4I0HAGf1q1XXa51VAIb3fnp3bdGbA). I had left my dog Ben (who used to fly in the chopper with me mustering sheep and goats, as per photos in my previous books) out there, holding them near the yards, I hoped. It wasn't that he couldn't keep them together by himself, but I thought there was a chance he might get thirsty or lose interest without my continued presence. It was quite conceivable that we could grab all the gear and rush out there to find no goats and Ben asleep under a tree.

It's worth mentioning that feral goats mostly roam unrestricted through the pastoral rangelands and can be there one day and gone the next, so when you spot them, you need to grab the bull by the horns, so to speak, and quickly. We needed to get some goat panels out there ASAP to attempt to get them yarded up and contained. It was worth a cash bonus if we could make it happen.

After Herbie reminded me of that story from four decades earlier, I remembered that when I returned to the mob of goats, Ben had them all together, waiting for my return in the aircraft. The blokes eventually came out with panels and equipment; we yarded up and proceeded to truck the goats to market.

I'm sure I probably kissed Ben: He had kept them together by himself for over an hour without losing any. You can understand how that bond builds between you and your working dog.

When I get together with John and Herbie, in no time, the names and places start to flow back into my memory as they spin yarns and share experiences that they talk of with laughter and enjoyment. I can nearly taste these experiences; they are so real. And even the awful bloody experiences are coated in amusement. No sense in getting down about them. May as well try and see the funny side.

Like me, both men have seen the highs and lows of livestock commodity prices, the floods and the droughts, the best and worst of people, and the travesties of people and families forced off the land, or shared in the celebration of record prices and good seasons, which rarely seem to come together. In many areas around WA, there have been good years from 2020–22, but this is exceptional, although great for the economies and communities. These blokes are the sort of bushmen you want to run into if you wish to taste the bush or if you ever need a hand.

CHAPTER 3

The Pink & Greys in the Bin

The very large workshop complex's corrugated iron roof displayed large letters painted in white: 'CURBUR,' on a green background, to help wayward aircraft navigate from the air or RFDS (Royal Flying Doctor Service) aircraft confirm the correct location before landing.

In the saddle room, my favourite spot, there was even old camel-harness equipment from a bygone era; I remember the four camels that wandered around close to the homestead when I was a child. They were once part of a team from the early 50s and they were a bit scary for a little person if you were on foot. They were used for transporting wool south or backloading with groceries or supplies up along the Mullewa to Roebourne stock route, no doubt.

The route was also known as the Grealdine Mine near the Galena bridge crossing to De Grey stock route, opened by ET Hooley in 1866, which ran through the eastern side of Curbur and into the Milly Milly open block. I can remember as a very young child noting the sheep pads cut into the coffee rock along the stock route 30 or 40 mm deep from huge numbers of sheep on the drove. Hooley was also responsible for building wells for stock water the drovers could use along the route. As per Wikipedia: He drove his first mob of 1,945 sheep up the Murchison River along the stock route named after him on the 27 May 1866, then north through the watersheds of the Murchison, Gascoyne, Ashburton and Fortescue Rivers, arriving at the Fortescue after a journey of around three months.

What also amazed me from his diaries was how at one stage he rode over 1100 kilometres on his horse from the Ashburton River in the Pilbara, to Geraldton in three weeks to collect some fresh horses and take them back to the Ashburton where he had left his wife and daughter. That's at an average of over 230 miles (not kilometres) a week on his horse overland and across country and no roads in sight. And how tough and foot sure was that horse to be averaging over 50 kilometres per day non-stop? By all accounts, Hooley was a very good horseman.

It was one of those hot mid-summer afternoons at the Curbur workshop, from where I could look out on its western side to the horse paddock. I was probably five when I used to spend a fair bit of time around the workshop, after the workmen left for the day to go out on the run. I would watch them pack up their vehicle with equipment to go fix a windmill or do some fencing, or even put the crib (lunch bag) on their motorbikes for mustering livestock or checking that windmills were pumping for stock water.

After they left, the team of stock horses would usually come in a short distance from the workshop area, to feed on the chaff and oat mixes they were usually given by the stockmen responsible before they went out.

The heat was shimmering across the small paddock by this time of mid-morning and the ground was starting to heat up. The horses had just come out of the holding paddock, had a drink at the corner mill trough and padded over towards the feed bins spread around the holding yard, which they knew had been topped up.

The feed would come from the southern side of the large workshop building where the chaff shed was. Alongside that was the saddle room, filled with a dozen or more stock saddles perched on three-by-two timbers, one above the other in four racks up to three metres high, sometimes with saddle cloth underneath them and selected bridles lying over the saddle.

A station saddle room exhibit at the Murchison Museum (photograph by Marianne Humair)

The horses padded up quite rapidly in single file from the trough to the feed bins 100 metres away, but didn't exert any more energy than was necessary, as the hot sun was drawing from their every step.

Pink and grey cockatoos, or galahs, fluttered from the stock horses' feed bins in clouds, and didn't seem to be bothered by the hot ground that I couldn't walk on with bare feet. You had to run from shade to shade, because nobody wore shoes. The squawk of the odd crow would break the intensity and mirage of heat as they signalled to one another if there were any meatier snacks than horse feed.

Topknot pigeons hung around in flocks near the horses' feet too and came and went in droves whistling in flight as they do and dashing around the horses and bins to catch a skerrick of grain before the galahs. They mainly kept together and gave their little 'whoop whoop' noises of contentment, along with the occasional fanning of their tail feathers and little love dances to each other.

The crested or topknot pigeon (picture supplied by Marianne Humair)

When the mustering season was in progress for cattle or sheep camps, up to ten or fifteen horses would be in a mob, as the worked horses would take their days off together. They leant down over their feed bins, necks extended into the 44-gallon drums that had been cut in half around the drum; just the right size for a horse's head to have a bit of room without birds being able to get in. Not that they would have, anyway – that would have meant awful peril with a horse nearby.

The pink and grey galahs and the crested pigeons, with their upright little crests, gathered around each bin in anxious anticipation. The dry chaff and oats occasionally got into the horses' nostrils, causing a colossal snort that would blow tiny morsels over the top and onto the dusty soil around the container, and the collected birds would raise a metre or two off the ground in a cloud and then settle down closer to the nearby titbits.

Luckily for the birds, there was often the potential for feed at both ends of the horse, as, eventually, a tail would lift, and manure pads would splot onto the dry earth. But it's fair to say the tucker from that end was mostly left to dry in the sun and parched earth, and birds largely ignored them until the grain and seeds they were chasing had dwindled amongst the dust.

Not far from where those plops landed were those two back feet that would occasionally kick out like a bullet, sending a wayward bird into pigeon heaven. But in fairness to the horses, not many did that; it would only happen if an over-exuberant bird got too excited about a flying morsel and brushed the horse's back leg. Mostly, they would lift their legs as a warning sign or stamp their front feet to give the message that the birds were getting too close.

A swishing tail would lift the little flock of pigeons or galahs from the back end into a hover and then they would descend again just a little further away. The horse wasn't doing it to move the birds so much, but rather the pesky bloody flies that hung around in swarms and too would settle back down to their diet of horse sweat and other delights once the horse's tail had come back to the centre and was unmoving again.

There was also the odd mudlark, or 'Beeleep', as my brother and I used to call them as kids, from the noise they would chirp up with. These black-and-white birds would be on wide duty, picking up the odd fragment of chaff or insects fleeing from swishing tails. Mudlarks make the most robust little mud nest and would make it stick anywhere, the corner of a shed or on a windmill tower or platform somewhere. They make a shrieking noise for no other reason, it seems, than to let everybody within a one-kilometre radius know that they're on duty and consider themselves in charge.

While they patrolled the outer perimeter for morsels, the flocks of galahs and pigeons stayed close together in groups of twenty or thirty, depending on food availability. If you watched closely, there seemed to be unwritten rules that both horse and bird parties abided by, although things could rapidly change, mainly in the horse's favour.

For example, all the birds seemed to understand that, provided the horse was relaxed and standing still, especially if they were eating contently with their head practically out of sight in a bin, every bit of spare grain, even close to the horse's feet, was fair game.

But if, on the other hand, the horse lifted its head from the bin and, in an irritated fashion, laid its ears back and thrust forward with its mouth in a menacing scowl, it was categorically time for birds to move further away, or else potentially suffer the consequences. The message was quite clear.

You could see which horses tolerated birds better than others, as the distance between bird and horse was minimal. Older horses had seen it all before and were more prepared to share the snorted morsels, or perhaps after some heavy days of mustering, their aching old bones just didn't need any extra stress.

And my God, if you had an old horse whose teeth were going, or needed rasping, and it dropped large amounts of semi-chewed chaff from its mouth onto the dirt outside the bin –weeeelllll hell, that was bird Christmas. Shrewd stockmen used to look for these signs that a horse wasn't getting the feed's total benefit and bring out the horse gag and teeth rasp accordingly. Then they would apply the gag to lock and hold the horse's mouth open and rasp the burr off with a metal rasp.

As a child, I would be waiting for a bird to get crunched, as it seemed some of the older horses would lift their hoof to remove a fly or stinging ant and then stamp down. Because those birds had worked that relationship close to the bone, they barely moved from the outline of the hoof as it came down in a hurry to potentially remove the biting critter. Sometimes, it seemed there was hardly any distance between a hairy fetlock and a feathery bird.

But the birds wouldn't take a chance like that with a young, freshly broken horse who was new to this hand-feeding caper and had spent the last two or three years as a brumby, fending for himself on the open rangeland. There would never have been a reason for birds to get so close out there. These horses had rapid reflexes, and it took a while to get used to this attention from birds wanting to share their feed.

There was an old metal trough near the centre of the horse-feeding area which was strapped up into the air, around three or four metres long and a metre above the ground. I had difficulty looking into it up there at that height. I could nearly walk under it without bending my head. It was wired strongly to large mulga trees at either end and could feed three or four horses on either side at once. They had the comfort of the shade of the trees while they ate. It seemed to me some old stockman many years before had wanted his horses to eat comfortably.

There was a little worn-out hollow under this long old trough, where the front legs of horses over many years had stamped away unwanted stinging insects or where youngsters pawed with a bit of excitement when they first got fed there.

I remember Dad's old bay horse, Sergeant, wouldn't share much of that trough if it had food in it. Any horses in too close proximity, whether on the side where he was eating or not, got the ears back, raised mouth reaction and a specific horse message to get out of the way or suffer the consequences. He was a cranky old soul.

Sergeant had been around for a long while, and his seniority was stamped on his movement around the mob of horses in the large holding yard. At feed time, when bins were topped up, he would make a loop checking to see which had the best concentration of oats or grain. He would move from one bin to the other, still chewing its proceeds, lift his head, and lower his ears back against his head to the unfortunate customer at the next bin, who had to scurry out of Sergeant's way quick smart or suffer the consequences. It sometimes took a bite on the rump of the unsuspecting casualty, who may have had its head in the bin and not seen the boss horse coming. You would hear the noise of teeth dragging over hide, and then his teeth smashing together along an unsuspecting rump, with a startled squeal from the inflicted animal. A streak of hide 50 mm long would be removed in the process, and Sergeant had his way again. He did, however, have a few distant mates he allowed to graze closer to him in the long bin under the mulgas, although only once he neared his fill.

I would spend hours whiling away the time down at the horses' feed area as a child, watching the intermingling of species down the food chain, probably also beginning to understand how horses thought and reacted to things as I got older to ride them. I am pleased to have experienced that bygone era as a child. The days often seemed hot, but there was always time for watching, learning and daydreaming. That's what kids are meant to do, isn't it? Play with their own shadows, as they learn about the world.

CHAPTER 4

Cattle Muster and the Curbur Cattle Yards

Building yards was quite a skilled art, especially as, more often than not, an adze and sharp axe were the only tools available. One was lucky to drill holes by hand with a brace and a large wood bit that had been sharpened on a stone. The old timber cattle yards near the Curbur homestead were about 2 metres high and built tall and strong in times when wild cattle were pretty feral and challenging to handle. The Curbur yards were built by Sonny McCarley, a drover, in around 1948. I learnt to ride a pony in those yards. The large wooden round yard was also used for roping, breaking, branding and castrating young colts.

There was a wooden raceway, about 20 metres long, leading to a concrete loading ramp with a rail bull-hitched across its top at every upright. These braces were essential to stop the race from spreading apart if bulls or other livestock tried to turn around in it. Similarly, the wide 4-metre gateways were braced with bowed or arched timber rails or beams, making them strong enough to hold the heavy gates without sagging when they swung open and shut. The arch needed to be high enough for a mounted rider on a horse to duck under without risk of decapitation or the loss of a hat. Quite important, that.

A story was related to me recently by somebody who was in the main receiving yard with a mob of cattle while marking one day in 1950. A young jackeroo had just purchased a brand-new pair of R M Williams jeans, which, in those days, were made to last through any conditions. The kid was sitting on the top rail, watching the old hands go about the business of marking and handling cattle, when this pointed-horn micky bull charged the rail the kid was on. One horn lodged deeply into his brand-new jeans, and the micky proceeded to take off, the kid and his very strong jeans firmly attached to the horn. The story goes that the kid was so scared, and had such a good grip on the rail, that he managed to rip the rail right out of the wooden yard panel and onto the ground, much to the astonishment of both the micky and the kid.

Now, he must have had a serious grip, because those mulga rails were wedged into the upright posts at either end, as well as bull-hitched with double eight-gauge wire. The micky eventually managed to dislodge his incredulous passenger, to the accompaniment of much raucous laughter from the other stockmen.

Back to the Curbur yards. The uprights were mostly made of ghost gums from the Curbur Lake, as the timber was softer to work with, although it wasn't sustainable to use over the long term, as it dried and rotted over time. Still, it lasted thirty or forty years; many would say that was long term.

Young men on horses relished the 'chase' of bringing scrubbers or breakaway cattle back into the mob for their fellow stockmen to see so they could get bragging rights around the campfire and show their standard of stockmanship. The wilder the cattle, the better the man it took to get them back, and the campfire chatter and legends grew accordingly.

Driving big mobs of cattle long distances to these yards was standard practice back in the 60s, but later, we built more permanent steel yards out in the country where the cattle lived and roamed, taking the road trains to them for better access on better roads. These stock trucks are up to two, three or four stock trailers hooked together, often double-deckers pulled by a single prime-mover truck. They are equipped to carry hundreds of cattle per load. This is a considerable difference from droving large mobs of cattle long distances to the house yards in that time.

I distinctly remember, as a little kid, hearing the constant bellow of small calves for their mothers and the continual crack of stockwhips when the mustering/droving teams were coming in from the west, nearing the homestead in the late afternoon before yarding up. The good old mother cows would turn back to their calves, calling to them, willing the tired youngsters to catch up. It was often the culmination of weeks of work for the stockmen, who had travelled considerable distances on horseback.

To a little kid, it would be a huge, exciting, mobile mass of noise and action when they finally arrived, hours from when you first heard that faint crack of a stockwhip and the bellows of the moving herd, way off in the distance. Then, as the sun was setting, they would finally appear in a cloud of dust, accompanied by the familiar ruckus of sound.

Horse riders would be on the tail of the big mob of hundreds of cattle slowly meandering along, some more reluctantly than others, calling back to their baby calves dribbling along on the tail of the mob, drooling and panting to cool off, in front of tired horses and horsemen who just wanted to see the last beast in the yard.

But the riders were also on guard, ready for action, because they knew that if there was going to be a breakaway beast, this was the time – within sight of the yards with the different smells and surroundings – when it was most likely to happen. The lead cattle would start to block back and not have their usual free stride as they examined the Yamatji camp, where all the families lived, with clothes flapping on the barbed-wire clothes lines: no need for pegs when you use barbed wire.

Little kids would gather in small groups to watch the excitement. It was better than the drive-in cinema in Mullewa. This action-packed event was right in their backyard. Some women hadn't seen their menfolk for a few weeks and were keen to wave to them, welcoming them home.

The occasional crack of the stockwhips kept the tired tail reluctantly moving. It could be 50 to 80 metres across, requiring two or three horsemen back there with a couple of riders working a wing or wing-tail position, keeping the mob together. When the cattle were fresh and flightier, a lead horseman might have blocked back the lead cattle until they settled to a walk rather than the trot, so they wouldn't get 'hot and silly'. When most animals – but particularly wild cattle – run too far, their temperature goes up and they often become more difficult to handle and to settle down. In this condition, they can become rather paranoid and confrontational, can be seen dribbling with their tongues out, and intermittently charge horsemen, vehicles or anything that gets in their way. They are separated from their usual common sense. That's why you try to avoid running them too long before enforcing a walk to cool down.

They were getting close now, and although a few cattle had broken away, they had been shouldered back into the mob. The lead cattle were blocking up at the entrance to the yard. They were not keen to enter and Dad, on his old cattle horse, Swagman, an experienced bay gelding with a big blaze, walked up to the lead and eased in behind some quiet leader cows to apply some pressure – get them heading in so they would start a lead mob for the rest to follow.

Swaggie, as we called him, had a bit of a swagger about him. His walking motion used to cause the hobble chains to jiggle musically on his neck when he walked, making his own particular tune. He was a big, thickset, powerful horse, unlike the thoroughbred crosses that were often used for cattle work.

Dad, as the boss, used to curl his full-length whip in one big curl, which hung down near his boot on the off side of the horse, then into two or three small loops that he held in his right hand. Sometimes the big curl would get hooked slightly in his spurs or the heels of his riding boots. He often used to rest the wooden whip handle on his right hip, on the plaited leather knob at the end of the handle called the 'Turk's head'.

As a kid, I used a small bullwhip, which was much shorter at a few metres in length and didn't get all tangled up when you cracked it, unlike the full-length types.

The old cows hung back, even against Dad's extra pressure from the horse, and he didn't want to use his whip to turn the others back around, so he raised it in the air and shook it, telling the old girls to 'move it up'. Their understanding of the whip required no further explanation, and they hesitantly moved further in towards the large, square, receiving yard.

Dad retreated and proceeded to get another 'bite' of cattle to follow the lead of the cows he had just pushed in. They did, and the yard started to fill up. The rest of the crew blocked up the mob and any potential escapees with the odd stockwhip crack to hold the pressure on the rear of the mob, with an odd 'yaki', or noise, to keep pressuring the mob to move in.

Finally, all the mob were in the yards. As men dismounted from tired horses and shut the chained double gates, the horses shook their heads and the hobble chains on their necks ferociously, removing clouds of the constantly buzzing flies. Horse sweat and cigarette smoke, mingled with the fresh cattle manure smell around the gate, also attracted flies, against which the horses swished their tails in much aggravation. The noise of the bellowing cows, still unable to mother up (mothers calling for their calves) was deafening. It made the boys' conversation impossible to hear but blended into a normal commotion of cattle muster and a yarded mob. This noise usually meant a good mob, which was always a good sign.

CHAPTER 5

Hobble Chains Jiggle on the Neck

An old stockman told me a story that unfolded on the Oakover River a few years back, and how they mustered the scrubbers (or desert cattle) that survive out there in no man's land for months on end when there's surface water available.

In this environment, the Australian outback, a working stock horse often carried a set of hobble chains on its neck. These consisted of two leather straps or cuffs to go around each front fetlock and a chain of high-tensile round steel links or rings, with a swivel in the centre between the straps. These restricted the length of the horse's stride and were used to stop the horses from grazing too far away overnight or at dinner camp. This meant the horses could roam to feed in the natural surroundings rather than requiring hand feeding, which was often unavailable in the bush. The hobble chains were carried on the horses' necks, strap end to buckle, when not in use, and the steel hobble links rattled against each other, causing them to jiggle or 'sing'. Each horse's individual gait would create a unique song, and one could recognise the approaching animal by the sound of its hobble chain singing.

As this book has an international audience beyond Australia, I feel it necessary to give readers some background information so they can appreciate what happens in this outback environment, which even most Australians don't experience on a day-to-day basis.

For example, many rivers in the world, such as the Thames in London, have water in them constantly or all year round. In Australia, especially on the west coast, many of the rivers are dry much of the year and only run high and fast with cyclonic rains, mostly in the summer. So, to go driving across or camping in a dry river bed is par for the course in that environment.

This gap in understanding was brought home to me during a recent phone interview on ABC National Australia, when doing pre-interview discussions with broadcaster Chris Bullock, who has done no doubt thousands of book interviews in his years with the Australian broadcaster. He made the point that he, in fact, hadn't realised that some feral cattle in the pastoral rangelands do run away from the capture of the stockmen trying to muster them together. And it made me realise that this environment in outback Australia is actually completely foreign to even many Australians, certainly to those not in the regions.

But don't fear, because this is one of the main reasons I write as I do, to try and explain to those not privy to the Australian bush what happens out here.

It's actually a place where many, if not most, of our forefathers started in this country. And it wasn't just gold mining. Many have fathers, grandfathers, uncles or even great grandfathers who were shearers, drovers, stockmen, jackeroos, horse breakers or station managers, or, in later days, water borers for mining companies or bus drivers in the regions. Many have heard stories over the kitchen table since childhood which stimulate their imagination to find out more about the Australian bush and its people, and how the livestock survive there. And as one who has grown up born and bred in that environment, I want to make sure that people hear and read a true and correct record of what it's about out here, because this is a dying way of life in its original form and needs correct placement into history.

I've come to the realisation that so much of this bush stuff needs to be explained so people can completely grasp and understand it.

And that's not a criticism, it's just a fact. If you haven't been there or experienced it, how would you be expected to know? Like me going to the South Pole; I wouldn't know what to expect.

In the upcoming story, Jack and his horse Bess were to trap desert cattle onto the station property lease behind wire (control them) and then muster and truck them to market for many hundreds of thousands of dollars in cash flow, which would not have been available to them otherwise. Quite an amazing feat in anybody's language. Like winning the lotto, but busting your gut to make it happen.

What's notable here isn't only the years of skills the cattlemen acquired to achieve that goal but the amazing natural toughness of the desert cattle with their capacity to survive and the ability of understanding, passed down into later generations, to know and understand the way back to river pools that their mothers and grandmothers had taught them in generations before. The only thing that restricts these feral desert cattle from travelling back many kilometres into the fresh ungrazed desert country is surface water to survive. From those station property leases that adjoin the huge areas of desert or arid rangelands, there is a regular flow of cattle that run out onto the desert in the wintertime and are forced to return in the summer when springs or surface water out there dry up. They follow the creeks and river systems out hundreds of kilometres towards Lake Disappointment and the Canning Stock Route along rivers like Savoury Creek, which is only one of three rivers in Western Australia that runs east, towards the centre of the continent, instead of west to the coast.

These desert cattle have made this journey for generations and this is why it's so important to harness this experience in the generations. Keep your older cows without culling for age alone and foster this experience rather than remove the cycle of learning from future generations. Calves follow their mothers into distant desert soaks which few know of, providing a haven for a pastoral herd of cattle in tough drought times. This is hard, cold, experience that no amount of money can ever buy. This is called survival.

The cunning of a cow escaping with her calf from a muster,

although it is the curse of every stockman, is showing her calf how to survive. None of us are much good dead. And that calf in turn teaches its calf, and so the genes continue down the generations.

And in ten years, there's 'a mob of scrubbers' in the Savoury Creek area avoiding muster. Who says animals are dumb? Even the Man from Snowy River wouldn't see their dust. They would see, hear and smell him before he even made his first cough. Today's helicopter mustering pilots are battling to get them into the yard with such a manmade advantage and still the few get away to fulfil the band of scrubbers that have won and continue the genes of survivors.

My point is not that people are foolish or lack thought, it's just that many of these experiences are way beyond their comprehension. I mean, how many people today can imagine riding hundreds of kilometres on a horse to do a job. Hullo??

So please enjoy the following story, and I hope I have explained most of what happens for all my readers whatever their background. There is also a glossary at the start of the book for more explanations of bush terms and bush slang.

The little three-year-old chestnut mare was tracking down the cattle pad for Jack on a loose rein. Tiny dust eddies curled up behind her on the warm summer's day. She was a beautiful little horse, barely fifteen hands at the shoulder, not big or strong. However, she would give her best all day, and not many could keep up with her when she was dealing with a flighty mob in the scrub. She was just all heart. She would keep on giving until she dropped, and, when riding her, it was apparent that she just loved her work.

Jack had a pair of taped spurs on the back of his old worn-out RMs (R M Williams made boots where the tape actually covered the point of the metal spurs which minimised impact on the horse). However, Jack only used them on the odd occasion, such as when he needed to get up swiftly alongside an escaping beast to shoulder it back into the mob. Only then would he maybe give her just a little tickle. Then she knew it was urgent, and he would feel her gather herself under him to give him that quick injection of speed.

They had developed magic teamwork over the time they had been together.

When she would get that excited gait going, especially along a well-worn cattle pad smelling cattle ahead, she would make the loose chains on his old spurs rattle. That was her way of saying, 'I'm ready.'

Jack had liked her from the first time he had run her in the round yard to break in before cattle mustering camp two years earlier. She had been the only chestnut of the three wild, unbroken horses that had been run into the round yard together. They had tried for just the two new horses into the round yard, but the four-year-old colt had forced his way in behind, beating the response from the young jackeroo on the gate, who just hadn't got it yet about wild bush horses and their turn of speed. And how quick and smart they were to weigh up a situation.

The round yard usually had only a few gates leading from it. Today, this is often called a 'pound', practically a round yard made of five or six gates into a separate yard each, giving you more drafting ability.

Jack had taken the pressure off the three horses in the big round yard and stepped back to see how they responded to this first interaction with scary humans. While the other two had been climbing over each other's necks, looking for an out, hindquarters to Jack, whacking their heads into each other and into the rails they were inspecting for an escape route, she had faced him, giving him her full attention. There was a lather of sweat and foam on the necks of the colts where their bodies had crossed over each other in the flight response away from this two-legged predator, who obviously meant to harm them in these frightening new surroundings.

But the chestnut filly, who Jack was to immediately name Bess, faced him with her near side leg and shoulder quivering with slight trepidation. Jack guessed that she was just over two years old, younger than the other two. She shook her head a few times in a nodding yes-type motion as if to say, 'Well, I'm ready to check you out. Let's see what you've got!' Her body language was in complete contrast to her two mates, the bald-faced younger bay colt (slang horseman's term for a horse with a big, broad, white blaze down its nose), and the larger, more muscular, five-year-old dark bay colt.

Jack would later get her in the round yard by herself once he could draft her separately. Still, there were approximately fifty-odd horses in the big yard they had run in that morning, all ready to be selected: around four or five for each stockman and a few spares left over for the mustering camp string.

Jack thought he had better take the draft gate on the pound or round yard to select Bess from the other two colts into his yard of selections, as he had a couple of nice types in there already and didn't want either of the two colts getting in there and mixing with his choices, which thus far, all seemed to have sound temperaments. Not that they couldn't be cut out later, but he would prefer it if they didn't stir his selections up.

This horse selection was the most crucial choice a rider made for the year because, although it's possible to knock some rough spots off a horse, it's so much easier if the animal is working with you. An animal that wants to buck, pull, shy or display other negative traits can make for a long day of mustering.

On further consideration, Jack decided he would have a better look at Bess in the round yard by herself later, but that wasn't possible just yet. He tried to set it up for Bess to be ready for the gate into his yard, to make it easy for the jackeroo to open the gate on his call, just for her to exit and not the other two colts. That task took some time, enough for Reggie, one of the older Aboriginal stockmen, to roll a smoke. Eventually, Bess was facing the right way away from the other two, and Jack called to open the gate.

As she galloped into the yard, he noted that, although she had held back and was steady in the round yard, she had a ton of 'get up and go' and was through the gate before the jackeroo even realised it. Jack licked his lips in anticipation as he noted her four white socks; better shoeing, because a light-coloured hoof was slightly softer than darker ones. This meant a shoeing nail could be driven in more easily.

But that had all been a few years back. Today was different.

There was no immediate pressure on horse or rider, as Jack went to check a section of the country where some scrubber cattle came out of the desert on the eastern side of the out-station in early summer, when the surface water pools had dried up out there.

And there was a new addition: A six-wire boundary fence had been built for the first time, blocking the scrubbers from coming into water holes on the river on the station proper. In Jack's experienced estimation, the scrubbers had been held away from the water for a couple of days, as October was getting warmer, towards 35 degrees centigrade. The fence had been erected by the new fencing contractor on the station. It had barbed wire on the top and more halfway down; a 'brisket barb' with four plain wires and 'steelies' only five metres apart for this section, which left the fence vulnerable to pressure.

Feral desert cattle wander around hundreds of square kilometres in their smaller groups, mainly through the cooler months of the year. They don't need much water until the summer temperature reaches 30 degrees. This is especially true if it has been a good season in the desert country and, because it has not been heavily stocked, the vegetation is in good condition. Even green spinifex contains a high moisture content, enabling desert cattle, hardened from years of living under harsh conditions, to survive.

As the surface water recedes and dries up in the Crown land desert country, the mobs of cattle are forced to travel back to station lease country, where graziers supply their herds with water year-round via permanent or semi-permanent river pools (sometimes spring-fed), windmills or water bore pumps.

Stockmen want a suitable type of quiet cattle who spend their energies feeding, producing calves and fattening rather than trying to run away and escape capture. However, these domestic animals are often at the mercy of the environment.

Feral cattle have survival in their DNA. These desert cattle can smell water for kilometres; their complex noses can seek out water like a cat does a mouse. It's heartlessly called 'survival of the fittest', but that is life in that environment. A formidable sense of survival is not a bad hereditary trait to include in your station herd if you can reject some negative traits like poor temperament, with which it often seems to run hand in glove.

Old cows or scrub bulls can remember following their mums to the same unfenced river pools that last the longest, perhaps many years before. It's like their natural GPS, where the battery never runs flat and there's always a signal.

These desert cattle have never seen men; most of them have little respect for, or understanding of, yard- and fence-building materials. Despite having a natural mobbing-up tendency with their mates, when push comes to shove and their survival fear overcomes them or they feel their freedom is threatened, they can cause many difficulties. These include safety hazards to staff and damage to handling facilities. It's crucial that young staff are trained to understand the potential danger, but even experienced cattlemen have been seriously injured. Feral, uneducated cattle can be very difficult to handle and process compared to regularly handled herd cattle from the station. They are well known for charging men, vehicles, horsemen, yard infrastructure and anything else that stops them from returning to the safety of their home rangelands. The faster and wilder cattle push to the front of the mob of escaping cattle and are the leaders, heading wherever they want to flee to, rather than where the cattlemen want to drive them.

Jack and Bess arrived at the boundary fence, checking out the 'trap entrance' which had been constructed to trap the desert cattle inside the wire as they advanced further down the river, looking for drinking water as pools dried up.

As Jack approached, he could feel Bess's stride quicken. She could smell cattle and the potential for excitement associated with the need to swing the lead of a mob of wild cattle. The hobbles rattled on her neck in a quicker frequency and the dust eddies grew thicker as she hurried towards the boundary fence line, hooves sinking into the softer river-plain dust on the well-used cattle pad.

As they approached the point where the new boundary fence had been erected, Jack eased Bess's excitement with a slight rein. This was not a usual muster, just an inspection. He let her know she could ease back, and she immediately steadied her pace. Jack positioned Bess on the northern bank of the Oakover River, away from the gutters that ran into it along the path used by generations of cattle to get into and out of the water pools. They know where the best going is, and cattle pads cut deep into steep banks, like nature's Main Roads Department.

Cattle or livestock wishing to travel parallel to the river for feed must cross this gutter (the steep-sided creek which runs into the river; which can be quite a few metres deep), which would be impossible to access without these V-shaped cattle or livestock pads cut deep into the sides of the drain, up to four metres deep. These pads were created from many years of hard hooves chipping them away, allowing quick single-file access into and out of the gutter and, hence, travel for livestock along the riverbank.

From his vantage point, Jack could readily catch glimpses of the riverbed and its sandy bottom, as well as the river crossing (a fence that goes across a river or large creek to stop livestock from accessing it when it's dry). It seemed to be in good repair; he had half expected some feral cattle to have pushed through it so they could head downstream to a more permanent waterhole in the river on the station lease. Their respect for constraints or manufactured fencing and restrictions was limited, and he knew that when it came to their sense of survival, not much would stop them. It was great to see no damage, because it saved him a job. He heard his fencing pliers and wire strainers rattle in the saddlebag at Bess's quickening gait.

But he and Bess could feel the tension from a large mob of scrub cattle off to their left, a little away from the river in the shade of an old ghost gum thicket. Jack guessed there must be 150-odd, as many were out of sight a few hundred metres away. They were restlessly lifting their heads from their peaceful afternoon slumber and raising their nostrils to get a better scent of the approaching foe. Usually, they would have galloped off by now, in a cloud of dust. Still, they moved gently and mooed to their calves in a stifled fashion as he pulled Bess to a gentle halt in the shade of a gigantic old river gum.

They collectively wanted something: the life-giving resource at his back. They knew it was there. They had drunk their fill from it in years past and, as little calves at their mothers' heels, had tasted that cool, fresh, restorative water. Their fear was now overcome by their need to survive. They would see what this stranger had in store for them before they fled, and anyhow, he was still a fair distance away.

He could see some of them were hollow in their flanks; two or three days without water, he guessed. Not so much of a concern with these tough boys and girls, who were used to going without water for long periods. And many were hardly thirsty, with flanks fattened and gleaming from months on ungrazed desert herbage, so they were probably only starting to look for drinking water now.

Both rider and horse were amazed at the mob not galloping off, but instinctively knew what was holding them like glue in the dusty, shady surroundings for the time being. It was their collective desire for water.

The fence line's V-shaped entrance into the station lease was approximately 50 metres off the northern riverbank; the initial gap in the fence was 80 metres wide at its entrance. This funnelled down into a metre-wide entrance gate that could be panelled off when required to stop cattle returning to the desert. And as this new fence line was 10 to 15 kilometres long, crossing the Oakover River at right angles and extending a considerable distance on either side, the station manager and other experienced stockmen in the camp believed that the desert cattle would be inclined to assemble around this location and could be funnelled through this trap gate to follow the ever-drying river pools further towards the station lease.

This was a trapping plan to relocate desert cattle back onto the station lease to be mustered and trucked along with the station herd cattle to settle down their out-of-town friends. This would put some serious bonus dollars back into the bank account after years of low rainfall. And there was usually a high percentage of scrubber bulls (uncastrated bulls), often weighing 600 to 800 kilos each, lifting the usual sales figures considerably. But you often had to earn every dollar with them, as you needed bull buggies and a bull cart to roll, hobble and load onto so they could be dropped into yards for transport to market. Once yarded, they were never let go.

Jack halted Bess in the shade, not wanting to disturb the cattle. He observed them as he made his rollie, not making any quick moves, and catching Bess when it seemed like she was going to give the itchy hobbles on her neck a damn good shake. Jack reached forward, gently removed the hobbles, and carefully hooked them over his saddle horn to stop her from shaking them. Something like that would be all that was needed to get the mob up and off camp, driving them back into the desert and towards Lake Disappointment, many kilometres away, with no guarantee that they would return.

After fifteen minutes or so, a few wary members of the mob had settled back down, and Jack had finished his smoke. Bess had relaxed too; she eased one hind foot onto its toe to rest, chewing on the metal bit, then pricked her ears backward to check out what her boss was doing. Jack eased her gently over to where the rope gate blocked the entrance to the station lease from the desert. He had made it himself from plaited rawhide, making it light to carry but strong enough to prevent a bull getting his head through the small square holes to push through to water and also to stop the cattle returning back through after they had drunk. It was designed to be removed so that the desert mob could come through eventually and replaced before they wanted to return to the desert after drinking. As he got closer, he could see over the fence the cattle tracks working the riverbank clay from the thirstier desert cattle, keen to access those river pools downstream, but held back by that new wire fence.

Jack was very much aware that the desert cattle were watching his every move from their shady retreat; he leaned down gently, undoing the 'cockies cotton' (wire) holding the rope gate to the wooden upright posts and box strut, which was designed to maintain the fence strength right to its end, so bulls couldn't knock it over whilst milling around and fighting amongst themselves and defeat the well-laid plan.

After removing the ties and gently placing the rope gate over his saddle, Jack eased Bess backwards for practical reasons and as a training exercise for her. He also purposely hadn't dismounted, because wild cattle are spooked at a man on foot. It's in their DNA, like a deer seeing a tiger ready to attack. It tends to startle them instantly. As he was slightly closer now and within clear sight of the mob, backing away took the pressure off the cattle, giving them less reason to startle. Jack knew that one beast feeling threatened and turning away in fright could start the whole mob off, which was the last thing he wanted. If they got a spook-up, it might be 100 kilometres before they pulled up.

From a training point of view, it was good for Bess to pass these little tests. He never knew when it might be a life-or-death situation for them to be able to accomplish tasks like these. Better safe than sorry, and this was an ideal training opportunity.

To Jack's satisfaction, there was hardly a movement from the desert cattle as he reversed away from them some 20 or 30 metres and then proceeded to hang the rope gate up in a river gum tree, out of harm's way.

Jack planned to leave the entrance open that night and return the following day to take another bite of desert cattle. He figured that since the closest river pool to the small entrance gate was approximately 5 kilometres downstream, it would take those cattle twenty-four hours to go down, drink and then return to the gate. They would settle after drinking for a while and mingle with some of the station herd cattle. The bulls would undoubtedly welcome the availability of new females, and they would not see any significant reason to return to that entrance right away.

He was also thinking that although they would remember coming through that narrow entrance, they might not be able to identify the small entrance point from the other side, especially when repelled by its shape on the station lease side of the fence. So, with experience and common sense on his side, he figured a day or so would be about right before returning to reset the rope trap gate.

Jack knew he had around 5 kilometres to ride back to the closest river pool where he had parked the 4WD and the horse float, and where that same desert cattle mob would drink. There, he could unsaddle Bess to make the 40-kilometre trip back to the homestead.

After the pressure of the situation earlier in the day with the mob of cattle and the satisfaction of a good job, Jack decided to stop in the dry riverbed and boil a billy from his chest bag for some crib (dinner). He dismounted for the first time in several hours and loosened Bess's girth so she could relax on the sand under the shade of a river gum. He thought about removing the saddle and letting her roll but figured that sand and sweat wouldn't be a good combination. He decided to leave it on until she dried off. She seemed happy just having his body weight off for the time being.

There was always plenty of kindling on riverbanks. He built a small billy fire, lying back against the soft sandy bank to wait for it to boil. Jack lowered his old worn hat over his eyes to keep the flies off and drifted into a doze.

A noise shook him from his slumber; he looked at his quart pot, which was boiling, giving him a timeframe on the length of his kip. But what was that peculiar noise he had heard? Looking across to the southern riverbank, he saw a dingo bitch with a few noisy pups, only a few weeks old, trotting towards the same downstream pool he was heading for. The noise that had woken him was the yapping of pups running alongside and terrorising their mother; she would have none of it, though. Although she had seen Bess, she had not seen or smelled Jack quietly lying against the riverbank; down in the sand, and the little campfire was also partially obscured from her view.

Now, dingoes and feral domestic dogs are not as troublesome in cattle country as in sheep country, however, they will readily kill calves and are still considered a feral pest that should be destroyed when seen. Jack wasn't carrying a rifle with him in the saddle; it was back in the truck. He could only watch as the bitch came to a standstill and Bess turned her head around 180 degrees at the movement and noise. The dog had probably got a whiff of Jack and the campfire by then. Jack smiled to himself as he saw nature doing its thing, way out here in the scrub.

It was only the bitch that saw him, as the pups were still playing, unaware of the dangers of the outside world. She was on full alert as she took the kids on a new road trip. She paused for a moment as if weighing up the danger to her family, then continued to trot on at an increased speed, much to the pups' dismay. However, their mother's increased speed must have hit certain natural alarm bells in them, because they stopped the yelping game and hurried to keep up with her in quiet obedience.

As she disappeared from sight on the opposite riverbank, Jack remembered the billy on the boil; he snatched a good handful of tea leaves out of his flour bag and a stick to quickly remove the billy from the heat to prevent it boiling over. Settling it down gently by the fireside, he firmly tapped it with the same stick to completely immerse the tea leaves and let them settle for a while. None of this swinging it around your head bullshit; he smiled to himself.

Bess shook her head and neck vigorously to remove a massive cloud of flies around her. The hobbles, which Jack had replaced, removed the last stubborn fly.

By the time Jack had nearly finished his panican (metal bush cup) of black tea and was almost ready to pack up his saddlebag, a mob of cattle were also heading down the cattle pad towards the same pool as the dingo and her pups. A dozen or so were station herd cattle, earmarked accordingly with notches out of the near side ears back and front; every property has its registered earmark and brand. A diamond out of the tip and a 'y' behind, he used to say to jackeroos or backpackers in the marking yard at cattle muster when marking the young calves. These cattle had all been handled and marked over the years, and they halted their single-file line on the cattle pad to look Jack and Bess up and down before quickly trotting on to the pool.

Once saddled and moving again, Jack decided to get to the southern bank of the river and follow the mob of cattle he had just seen on their way into the long, cool river pool. It seemed to be easier going on that side. Once Bess could smell the fresh scent on the cattle pad ahead of her, she started to speed up again, believing she was now into mustering mode. Jack gave momentary ease on the bit to take her out of overdrive and back to a speed that suited him.

When they arrived at the pool, the cattle before them had drunk and were resting in the shade as both horse and rider passed them to go to the 4WD and float, which were now out in the afternoon sunshine.

After giving Bess a cool drink at the river pool, Jack dismounted, his right boot landing lightly onto the dirt like a cat jumping from a table, cushioning his weight onto the ground, making it easier on the horse's back and not rubbing against the girth unnecessarily. It also works in reverse: you learn to get back into the saddle quickly, with your right leg close to the saddle as you swing in, so if your young mount is likely to pull away in fright as you mount up, you are closer and have better control of the horse. This may avoid a long walk home. Old habits die hard, Jack thought to himself, as his father had taught him better safe than sorry. He pulled the reins over her head; she desperately wanted a head scratch and for that bridle to be replaced by the simple bush halter on the vehicle tray.

Holding the bush halter rope after removing the saddle, he let her roll in the soft sand nearby. She took a few minutes to select a spot, walking around in a circle, semi-bent at the knees, until she crumpled with a huge groan and rolled over with all fours in the air, pushing her wither deep into the soft sand, getting both sides covered like a massage of natural luxury. She sighed as she stood up and shook everything bar the smallest bits of sand off herself. Jack threw the lead rope over her head and said, 'Get up.' She walked casually around to the rear of the float ramp and walked on with that particular sound of a shod horse on a wooden floor.

Jack smiled. He enjoyed that little trick, chuckling at the thought of dramas he had seen over the years when other riders tried to load their horses onto floats.

He lifted and latched the tailgate, tied her halter rope nice and loose, so she could reach around for the odd pesky fly bite, and drove the 40 kilometres home.

The Muster of the Scrubbers

Jack returned to the boundary trap gate a day later than he had intended. He was a bit worried that the scrubbers would have had enough time to go back to the area where they had come through the boundary fence, found the small exit hole and escaped with a belly full of water, back to who knows where.

But to his pleasant surprise, he noted a good number had come in, probably most of those he had seen under the trees a few days before and, although a few tracks near the small exit indicated that some had tried to find a way out, none seemed to have found it as yet.

Jack was on a bay gelding today, an older camp horse of five or six years, who was reliable but didn't have Bess's spirit or spark. He was OK, although Jack hadn't ridden him enough to teach him significantly.

He retrieved the rope gate from the tree and placed it on the station side of the exit point to give it more strength against those that tried it.

He had noticed a few scrub bulls he'd not seen before near the waterhole on the journey to the boundary. He presumed they would have accessed his trap gate to arrive there.

Surprisingly, he didn't see any cattle outside the boundary, although he noted a few tracks over the fence. He decided he wasn't going to go and disturb those cattle, which might be coming in. Most were wild and uneducated, and Jack would potentially do damage chasing them around, even if he managed to fluke a few through the tiny entrance trap gate. Better let them mooch around without any pressure and eventually find their way in like the others.

The plan was to give it another few days of trapping to trap maximum numbers before mustering this area inside the wire. Jack rode up a few times during that period to remove and then reapply the rope gate so that those on the outside were a bit thirstier and keen to examine entry points to the water as well as to block any that might be smart enough to find the small escape gate and make a habit of using it.

It wasn't a matter of deliberately letting cattle perish outside the fence, because nobody knew where they were coming from or when their pools in the desert might have dried up. There might be a few more weeks left in waterholes out there, but it was getting too hot to handle and truck cattle on the station if they left the muster any later.

The mustering aircraft was booked, and the horses, bikes, bull buggies and camping equipment were packed to camp near the pool which Jack had been using as a starting point.

It was before dawn on the first morning of cattle muster, and horses and motorbikes were being unloaded near the pool to await instructions from the two choppers, who were already heading out from the homestead to an area a few kilometres east of the boundary fence.

The plan was to send the horses upriver at once, because they didn't make any noise, but hold the bikes or buggies until given the all-clear from chopper pilots that cattle were inside the fence and engine noise wouldn't be a factor. The trap gate was already open, and cattle from outside could filter through, acting on the distant noise of the choppers. There was no direct pressure to confuse the cattle and get them stirred up and on the trot, just that faraway aircraft noise at a distance encouraging them to go back, the sound travelling many kilometres in the cool morning air.

The choppers reported to the ground crew that there were no cattle a few kilometres out from the boundary fence, but there had been a few tracks around in the last day or so as the chopper pilots could readily see from the air, and they kept on working west towards the fence with their patrols.

The chopper pilots knew about the trap gate, of course, and said they could see dust and cattle closer to the fence line ahead of them. They were purposely hanging back to give the desert cattle time to get through the tiny trap entrance, only designed for one animal to enter at a time, but also watching for mobs to start running north or south along the boundary fence on the eastern side, where they wouldn't be able to get through its new erection. Everything was conveyed to the ground crew from the air by UHF radio; each crew member had a set in a leather bag strapped to their body.

One of the choppers went to the southern side, some 3 or 4 kilometres south of the river, to help block that southern retreat along the wrong side of the fence line. This was also to start heading cattle on the station side of the fence line over the river to the north. The trucking yards they were mustering to were approximately 10 kilometres northwest of the trap gate, and ideally that was the general direction every animal was to head towards.

The plan seemed to be going well, with only fifteen or twenty jumping over the fence into the station lease just north of the river and trap gate. From the distant dust, it seemed like a hundred or so had been funnelled through the trap gate, where one side was partially damaged, meaning something must have gone through there in a hurry.

Jack and the other four horsemen were up near the trap gate by this time and ready to drop in on the tail of the main, or coacher, mob. As they were travelling quietly with their radio volumes low, they weren't likely to cut off or turn back any cattle. However, they had seen a few mobs ahead on the trot from the chopper noise heading north-north-west. Most of them were marked as station cattle, which had come from south of the river. The second chopper was only a kilometre south now and would put all the horsemen onto the rear or tail of the mob shortly, because that's where they would all enter the action initially. And then they would be perhaps called out, away from the tail to gather straggler cattle if required

Jack had decided to use Bess for the big muster; she was on the jog and was keen to go as soon as needed. When Jack had that radio on, she knew it meant a full day of action, and she was ready to rock those hobbles on her neck into a choir.

Just as the last mob on the tail came over the river from the chopper pressure, Jack knew they had their 'coachers', as he could see these were all quiet cows and calves. They were ideal for steadying up those scrubbers when they finally caught up with them, now possibly way up near the lead furthest from the chopper noise.

Because this last mob was behind the rest and already on the trot from the chopper, the horsemen chose to follow them at a steady trot.

This tail mob were going in the right direction, and then they could take them in hand and settle them down to a walk a little further on, as was usual with a coacher mob.

As all this was happening, the second chopper had gone north outside the boundary fence line. This was because all the scrubber cattle had gone either over the fence or into the station lease country by one means or another. And that helicopter was heading all those cattle in a more westerly direction, towards the yards, around 8 kilometres away.

Jack and the other horsemen's coacher mob were steadying now to a walk and the stockmen were starting to see other herd cattle just up ahead, which they would begin to pick up along the way, but there was still no sign of the cleanskin scrubbers everybody was looking for, not yet anyway.

Jack moved up on the top lead of the coacher mob, mobbing up with others in front. He could see an old scrub bull, around five or six years old, that had stopped running and pulled up in the shade. The bull was rich dark red and a longhorn on either side, with one horn turning down and the other up. He was fat and shiny; he had come from fresh country some distance away. Jack said over the radio, 'Take him along quietly until we get him out in the open off the river with a bull buggy. These are the ones we don't want to lose.' As the bull galloped off, blowing snot in Jack's direction, his body language was enough to let both parties know that this bloke would not go easy, and was very unhappy about the position he was finding himself in.

Compared to earlier times, the choppers do the bulk of the mustering work nowadays, but everybody enjoys progress. Even Jack, who had done it the hard way without any aircraft support many times, was happy to drop back behind the odd mob and follow the helicopter's pickups into the main mob. Instead of galloping like an idiot to turn the mob, the chopper had done all that, and he just needed to come in behind when the helicopter went back out for more stragglers. But horsemen still needed to be alert and have their wits about them for scrubbers or cattle that tried to break away to freedom. Time was crucial. If they missed the jump and were caught sleeping, once an animal was even 20 metres away from the coaches, it might be too late to get it back. Especially when they used the vegetation to help insulate their departure (as the cunning cattle do), waiting for and hiding in a band of thick scrub. This was their country, and they knew it well.

Jack was guessing that they probably had 1,200 or 1,500 head just in front of them. He and Bess dropped onto the tail of the coachers with the other horse riders, as the tail was starting to lag. Jack pulled out his whip to make a few cracks to liven up the rear; there was no need to hit any animals to apply that bit of pressure to keep the cattle walking. Even the three or four motorbikes that had joined the tail were not having much effect when it came to getting the tired tail cattle to move willingly.

It was only a few kilometres from the trucking yards now and the country was opening up, and the three bull buggies had joined the main mob. The most inexperienced driver was keeping up the tail and learning the ropes. The other two took one wing tail position each, because in case of cattle breaking out, that was likely where they would come from – top lead or wing.

All the ground crew were in radio contact with each other because they were all in close-range ground-to-ground, but the choppers were hanging further back now to help with the pending yarding up when needed with that extra noise and pressure.

Jack figured there must be nearly 2,000 cattle in hand now, grouping up as they got closer to the yard as the mob had come together, and he could see some of the scrubber bulls and wilder cleanskins starting to get a bit anxious from the close proximity and pressure of horsemen and vehicles.

The boss told them on the radio to back off and give the cattle some space. They were keeping it together so far, but some were moving towards the outside of the mob, ready to make a quick exit. The larger animals near the centre of the mob stood out because of their huge size, and they were keeping as far away as they could from human contact. There even looked to be some old piker bullocks, who no doubt had been away somewhere for years, but it wasn't clear which station property they belonged to yet; they seemed to have quite different earmarks.

'Look out!' a voice on the radio yelled as the scrubber bull that Jack had seen earlier in the day broke out of the mob behind the bull buggy on the right side. The driver wheeled his vehicle right hard lock, the steering wheel in a full clockwise direction as he heard the cry. The V8 howled whilst spinning its wheels, doing a 180-degree turn in the dust, and sped rapidly around the beast. As the animal came up to the passenger side door, it hooked and ripped with its mismatched horns and managed to get under the pipe bars on the 4WD, nearly tipping the vehicle over. The driver, although quite experienced, was taken aback for an instant, but he collected himself and again went after the animal, which had now cut in behind him. As the bull had gone out past the tail of the mob, the driver used his driver's side door to try to shoulder him back into the mob. No hydraulic arms here that come out from the front of the vehicles to capture the animal around the neck in a vice-like grip like on TV. Just the need to physically force the beast back towards the mob.

As the colossal beast ripped up against the driver's door, hooking and tearing with bellowing rage, his downturned horn slipped between the protection bars, just centimetres from the driver's leg. He had sensibly moved further from his foe and was now only sitting on the inside quarter of his seat to avoid those ripping horns.

In the meantime, a cleanskin cow had broken out of the other side of the mob in front of Bess and Jack, and it was on.

At the same time, three old bullocks had seen the opportunity to make a break; they were trotting away past the lead, sending the other bull buggy, driven by the manager, to try and wheel them back. 'Block up the mob', he yelled on the radio as he departed in a spray of dust. 'Block 'em up, block 'em up.'

The driver with half a bull's head wedged into his driver's door had turned the vehicle around so that the bull was closer to the mob, with his rump facing it. The animal continued to roar, mouth open and thick saliva dripping onto the leg of the driver, who didn't have much seat left. With one almighty wrench, the bull pulled back, bending the sidebars on the buggy, much to the relief of the driver, who had nearly run out of stamina, his adrenalin levels spent. Beaten and physically outmuscled, the bull returned to the companionship of the nearby mob.

Jack managed to get Bess onto the cleanskin cow's shoulder before she got too far from the mob, pressing Bess's shoulder against hers to physically force her back towards the mob. Jack added a cut from his bullwhip across her hindquarter to reinforce the message.

One chopper was helping the boss with the two bullocks who had split off from the leader and were going their own way. But after hooking at the chopper skids, the third of the old bullocks was using his horns to try and hook into the undercarriage of the chopper, a fear protection response from the animal. The pilot had managed to pull them both up in a bit of scrub cover. Helicopters have been pulled down physically and crashed by bulls or large cattle who have managed to catch their horns in helicopter landing or skid gear with their strong heads and horns. They can pull a small aircraft from the air very easily if they manage to hook a horn behind a skid, if the pilot flies too low and within reach, in trying to stop them escaping. What can start as a more threatening, protective head-thrusting stance from the beast which is trying to get away, can eventuate in a horn too close for comfort and an aircraft down.

The other buggy, departing from the rogue bull, now back in the mob, had come over to block them from reaching the scrub. As the chopper hovered above, the manager attempted to bring his old bullock back to the other two to settle them all down, but he had his hands full.

Meanwhile, the second chopper kept the main mob together; they were now only a kilometre from the yard.

Because of the proximity of the yards and the fact that the old bullock had ducked in behind him several times and had got a reasonable distance away, the manager decided to let him go; he didn't recognise the earmark on the old piker anyway. He returned to where the buggy and chopper were holding the two other bullocks, who by now had cooled off a bit in the scrub. It was arranged by radio

to take them back to the main mob. As the chopper came forward to move them from their scrub protection, with the bows of trees blown flat by the downwash, the old pikers tried one last separation at 45 degrees towards the mob. The drivers couldn't see that outcome through the scrub, but the pilot informed them, and each ran around towards where the old blokes were escaping into the open before steering them back towards the mob, physically using the vehicles. The pikers were swishing their old tails vehemently as they reluctantly headed towards their mates in the group.

Fortunately, with a mob that size, there were only a few more possible escapes between there and the yard, because there was a big mob of cattle for the scrubbers to hide in. This meant they didn't need too much direct pressure from the stockmen. The noise and pressure from the helicopters were very effective. Once the cattle were in the big round wing of those particular cattle yards, the way became narrow enough to put a line of vehicles and horses across to pretty much block off any potential exit. That's where stockwhips at the rear worked a treat, so a single horse and rider could go forward and keep taking cuts (taking bites or groups of cattle forward, so the others are inclined to follow) to move cattle into the large receiving yard. It's a big yard to hold 2,000 head of cattle, especially with some of the large bulls included. That would take some processing tomorrow.

It was an extremely successful muster, but only time would tell just how successful Jack's trap gate had been. Once they all cooled off and settled down early the next morning, they could be drafted, processed and counted.

Eventually, it was clear there were indeed a few hundred extra cleanskins in the yard, which was a big win for Jack and Bess. Jack thought a new pair of RMs to replace his old pair and an extra bag of oats for Bess wouldn't be a bad bonus.

CHAPTER 6

Some Old Bush Stories

'Dad, Dad, Chipper and Lassie are stuck together, down at the shed, back to back.'

I must have been around five or so when, one day, while playing with some stuff in the workshop, I realised that Chipper and Lassie were stuck together.

I had never seen this before. They were both moving around uncomfortably with what seemed like their backsides nearly touching, their tails partly raised in the air, and each dog seemed to want to pull away in the opposite direction from the other. I could see from my limited understanding of dogs that each dog's body language was very uncomfortable. And this concerned me greatly. On closer examination, and after unsuccessfully trying to pull the dogs apart, I realised I'd better get help.

Dad was having a PGA (Pastoralists and Graziers Association) meeting that day in the Curbur shade house, or coolhouse. This was a building designed with cane grass or spinifex walls where water trickled down its side and, when the breeze or air blew against it, it cooled the inside quite significantly. It's the Australian version of the waterbag effect. It was where the family slept at night in the summer heat (because there was no air conditioning then) or where you met when you had friends or neighbours to visit, as on this occasion.

I rushed into the middle of the meeting proceedings yelling quite loudly, still deeply concerned about the difficulty the dogs were going through.

'Dad, Dad,' I called above the meeting noise. 'Chipper and Lassie are stuck together tail to tail, what should I do?'

My father, who was always very quick on his feet in such circumstances, said, 'Go over to the wood heap and get the axe and cut them apart.' Huge, raucous laughter followed from the meeting of about twenty or so bushmen and women, who had seen it all before and had guessed precisely what was happening.

I knew Dad was joking about the axe from the smile on his face, and I would never be able to do that to our much-loved working dogs anyway. The roar of laughter from all those present caught me by surprise and embarrassed me, as I couldn't believe they weren't as concerned as I was. But I also couldn't believe that Chipper and Lassie had done something so wrong that we had to destroy them.

Maybe they were going to be like that permanently?

Fortunately, an older man beckoned me over and whispered in my ear that they would be OK: 'Just throw a bucket of cold water over them, and they'll be fine.'

I wasn't entirely convinced and was still very concerned, but on returning to the workshop noted that the dogs had now separated and were both resting under one of the parked 4WDs. I was very relieved, but still a bit baffled as to what had happened.

Chapped Lips

Bill, the young jackeroo, had only been in the camp a few days with Jack, the old stockman who was tough enough to make a four-inch nail quiver. But Bill had seen enough of the older man's skill and experience in the bush to realise there were lots of things he could learn from him. And despite the older man's peculiar habits in the bush, Bill had made up his mind, mostly because he knew he was as green as grass (he had already picked all these bush sayings up, see?), that he was going to make every attempt to learn everything he could from the old bushman.

After a week or so of riding fences on his old grey horse, out in the sun and wind all day, Bill started to suffer from dry and sore lips. Although Bill was not expecting Jack to have much medical experience on these issues, he figured he seemed to know everything else about things out here in the bush, so maybe he had a remedy for chapped, sore, dry lips.

As Bill was collecting some firewood to boil the billy for smoko, he thought it might be an ideal time to ask Jack the question about how he could improve his sore lips. Bill asked his question, and without saying a word, Jack went around directly behind his horse, lifted its tail, and kissed it straight on its you know where!

Bill was stunned and repulsed by the older man's actions. At first, he couldn't believe what he had seen with his own eyes.

'Jesus, Jack, that's bloody revolting. What does that do to your lips, moisten them or something, does it?' enquired the young man, still trying to come to terms with what he had just witnessed.

'Naaaaa,' said Jack. 'It doesn't moisten 'im, but it sure stops ya licken' 'im!'

The Old Yardman and His Dog

Yardmen were people employed on station properties to do odd jobs and chores around the homestead such as collecting firewood, watering the garden and butchering livestock for the table. They were once an essential part of the station but are no longer as common.

One old yardman had worked on a station out of Carnarvon for many years. When the new manager suggested that the old bloke take a holiday break after many years of service, the old man said, 'I wouldn't have taken a job on here if I thought it wasn't permanent.'

When the same old yardman went down to the killing pen, he just sharpened his knife on the steel. His dog, which he had trained and been with for years, would go out and muster the holding paddock for the killers – sheep set aside for eating on the station dinner table. Skilfully, the dog would bring them back quietly to the killing shed where the old yardman would select and catch one, and the butchering would be done.

Now the holding paddock was around 2.6 square kilometres. The paddock was designed to house the milking cow and some killers. And there was no saying whereabouts in that paddock those sheep were grazing when needed.

For a dog to understand in advance what task was required when his owner just walked towards the killing pen sharpening his knife, not saying a word, was exceptional. Knowing precisely what area was to be mustered and not getting his task mixed up needed incredible intelligence and understanding. On top of that, to know just where to bring the sheep to after he had found them and to muster them right outside the killing pen shed was a fantastic feat, especially all on his own.

While the dog was away doing his fantastic job, the old yardman would wash out the concrete floor and get all readied for the incoming sheep. Once the sheep were successfully mustered to the shed, the dog had to keep a tight hold on the mob, close enough to the shed door for the older man to run in and catch a selected animal or two.

Although there was a fence directly alongside the killing pen to hold the sheep against, once a single catch was made, the mob would split. Then it required time and considerable patience and skill from the dog to get the little group of ten or twelve back together again, as they would all run in different directions from the catcher's pressure.

In such circumstances, a younger man might be forgiven for catching two sheep at the one time. Not an easy task, of course, but done because he would not have had confidence in the dog being able to retrieve all the stragglers back into a mob again.

But the old yardman didn't have the slightest doubt that the dog would get the job done and so just caught one sheep at a time (which was difficult enough at his age, anyway). And then he awaited the dog to do his thing and remuster the stragglers back into a smaller mob to catch from. Both parties completely trusted each other to deliver, based on a unique and enduring relationship developed over many years.

Even more remarkable was when the big thumping 100-kilogramme merino wethers had just been shorn, with all the wool removed from around their eyes, making them even more skittish and difficult to catch by stealth. The talent necessary was a real art and required much more ability than just speed.

This fantastic relationship between man and dog continued for years very successfully and with no fuss. People just ate meat at the dinner table, not knowing how it got there.

But, as happens, the old yardman died of a heart attack one day in the homestead garden; the police and ambulance were called to remove his body and take it into town. His old dog was stunned and dumbfounded. He sat for days in the shade of the old gum tree near the garden, where he had last been with his old boss, awaiting his master's return.

Since the old yardman had passed, the station manager asked the jackeroos to kill the sheep until a replacement yardman was found.

And they figured it would be easy. A walk in the park, they thought. It would just be a matter of going down to the killing pen and sharpening the knife as they did so.

But when they did, the dog never moved for them; he didn't get up and barely paid attention, just laid his head mournfully on his paws and looked at them with big sad eyes.

And he never mustered the holding paddock again.

There was only one boss he did that for, and he wasn't around anymore.

CHAPTER 7

A Bush Boy Abroad

My mum had come to Curbur as a relatively young woman in the 50s and had been effectively cut off from her family and friends in South Australia. So, I think she was delighted that I was going out into the big wide world to discover new and exciting places, despite my young years. And from that isolated bush vantage point, everywhere was new and exciting.

In the little brown diary she gave me to take on my journey, this entry was her way of supporting my idea of going overseas to Europe.

It was something that she never had the opportunity to do herself. She and Dad had always talked about it: 'One day, when we go on our overseas trip...', although, when it came to it, I'm not sure Dad was really that interested.

'To Gregory, Remember, life is as you make it', Love Mother.

I left Perth Airport, heading for Greece, after having some goodbye drinks with Mum, my younger brother Keros and a group of friends. My father hadn't come; he didn't think travelling overseas was a worthwhile experience to celebrate. I appreciated everyone else making an effort to see me off, though. I think the drinks must have been cheaper at the airport in those days!

My old schoolmates, the Fortiadies brothers, were staying in Athens for some long holidays. If you stayed abroad for over twelve months before returning to Australia, you didn't have to pay import tax on luxury vehicles. Mr Fortiadies Senior, who'd lost several houses and other real estate in the horrific cyclone that devastated much of Darwin in 1974, wanted to buy a 450 SEL Mercedes with some of the insurance he had received. He had worked long and hard over the years and been planning this family holiday. The family was also there to give Michael (Hully) Fortiadies (a friend of mine from school) a holiday with all his Greek relatives, as he was going through some long-term health issues.

Since Athens could be a base for me from which to travel and work throughout Europe, and because I wasn't a seasoned traveller, I figured it was an excellent opportunity to spend a few months with them and get some European travel advice from them and their extended family.

Travelling abroad was a much more significant step in 1976 than today. There were no ATMs, only traveller's cheques, and keeping in touch was difficult. Mobile phones, text messages, Facebook and Skype were yet to be invented. There wasn't even email.

When I was in Athens, I felt I should ring home to let my family know I was OK because I hadn't been in touch for a few weeks. I could not get my traveller's cheque exchanged for cash, so I decided to ring reverse charge. International calls went down the lines from Athens to Perth and then from Perth to the Mullewa exchange, where the phone call had to be put through manually by the technician.

At the time, Kim Broad (Colin and Helen Broad's daughter, from Yalgoo, originally) was working at the exchange in Mullewa. I could hear Kim from my little phone box near Syntagma Square in the centre of Athens, telling my father that she had a reverse charge call from a Mr Keynes in Athens. Would he accept the call and pay the charges?

'No', was the response, and he put the phone down. The click of the Curbur phone came to me from the other side of the world, and I thought, 'Just as well I wasn't ringing from jail, Dad.' Kim returned to the line and said in a bemused voice, 'I'm sorry, the customer won't accept the call and pay the international charges.' At least I got to say hullo to Kim.

I knew Kim well, and we both knew what the tough old bush blokes were like.

One experience that shows just how green I was about international travel happened when I chose to stop at the Shangri La Hotel in Singapore en route to Greece. I was seriously proud of the black leather jacket I had bought for the big trip and had it on when considering taxi transport from the airport to the hotel. Big mistake, Greg, big mistake.

While everybody else seemed to be lining up for taxis in a long queue, I was getting knocked over by offers from drivers of limousines! It didn't take long to realise I should leave the coat home when I needed to buy anything so I didn't look like a tourist with more money than sense. Although, the limo trip was only slightly dearer than a standard taxi, and it was quite an experience.

That was my initial introduction to the real world of business in Asia.

After being blown away by the majesty and opulence of the Shangri La, I took my flight to Athens, where Hully met me. I tried to make sense of the Greek writing, but the thing that intrigued me during the trip was the panel damage that nearly every taxicab seemed to have. It took a few days in Athens to completely understand how they ended up like that. The road rules seemed to be pretty much a free-for-all. It took me some time to figure out who gave way to whom: No one gave way to anyone, and the dented panels resulted.

The massive Acropolis ruins, including the Parthenon, which sits on the city's skyline, are the first sight that strikes your heart when you enter the city of Athina (Athens). Unique history is staring you in the face as the imposing ruins look across the majestic metropolis. The mind boggles at what those ruins would have experienced in their time; they were built around 450 BC. The stories they could tell around a campfire...

Only our ancient Aboriginal heritage gives a more profound sense of history and time. But our 200-odd years of white settlement pales into insignificance compared to Europe.

Something that smacks you very hard in the face in Europe is just how young Australia is. We're barely out of nappies by contrast.

On the trip from the airport to the place where I would be staying with them, Hully took me to visit his uncle, Theo Christo, at his property just outside Athens. Theo Christo had been involved in defusing bombs on the island of Crete during World War II. A bomb had gone off and he was blind in both eyes; they were just empty sockets, with a black patch over one eye. He was also missing part of an ear, had considerable skin grafting all over his face and had lost one arm from the elbow. He used to chain-smoke, and his remaining hand had skin grafts and, apparently, not much feeling. The cigarettes would burn down to his fingertips, and I could smell his flesh cooking.

My little mate, Hully – he wasn't that small, actually, had quite a solid build and was a good footballer – for reasons only known to himself and his demons, had decided to disappear for two hours after dropping me off, without even mentioning that he was leaving.

Theo Christo spoke no English other than, 'OK, mate.' He had learned that from Aussie soldiers during the war. His wife, Coula, who was his eyes and ears, was also unable to speak a single word of English. Since my knowledge of Greek was zilch, you could say we had a very interesting few hours.

It's quite a cultural shock for an Aussie bush boy to sit in a house in rural Athens with a blind man and his wife, neither of whom speak English, and experience the constant smell of burning flesh. You must try and see the funny side, although it escaped me at the time.

I soon developed an arrangement with Theo Christo. When I said, 'Ella' (which seemed to be Greek for 'come') and tapped him on the elbow of the crook arm, without the bottom bit, it was a sign that it was time to light up another fag, because that one was starting to get a whiff of burning flesh again.

It was not the last time my little mate disappeared for a prolonged period. It was no good asking him where he had been when he finally returned, after many embarrassing hours. I knew I would not get a legitimate answer, just his mischievous grin.

I did stay and had a delightful time with the Fortiadies family in Athens; it was a marvellous base for me to use to build up more confidence to travel around Europe. When you walk out into a busy street in Europe, you quickly realise there's always somebody keen to help you spend your money and you are a long way from home. Welcome to Athens and Europe, Murchison boy.

The Van and Theo Christo

Little did I know that that challenging first meeting with Theo Christo would come in handy many months later, when his blindness, age and experience in Greek culture would save my bacon.

After a month or so abroad, I had the opportunity to buy a British Leyland campervan from an English tourist, complete with a homemade bed, stove and fridge. The price was right, but there was a hitch! In those days, when you owned a vehicle in Europe, you had it stamped onto your passport (not sure if this still happens), and you couldn't leave the country until you had the stamp officially removed.

This Pommie guy, Geoff, who I met while working on the charter boats at Piraeus Wharf, had to quickly get rid of his van because he needed to return to England in a hurry. He didn't have the time to go through all the transfer procedures, which were a major bureaucratic hurdle requiring lots of money in envelopes handed under tables. I witnessed this myself later, with a boat skipper attempting to get his boat registered in Greece.

The plan was to drive the van across the border into Yugoslavia at about 2 am, before the border guards changed, according to Geoff's mate, and get it transferred off his passport. Then, I would drive it back into Greece at about 3 or 4 am and get its registration stamped onto my passport and all would be sweet.

Yeah, right.

I knew nothing about the Federal People's Republic of Yugoslavia and was especially ignorant about how they resembled their Eastern Bloc brothers. The head border guard on the Yugoslavian border looked just like the brutal prison guard, Hamidou, played by Paul

L. Smith in the movie *Midnight Express*, a big, scary, mean mother. He gave you the very distinct impression he could throw you in a dark cell somewhere without a moment's thought. I think they look a bit scarier when you have a conscience about what you are doing, although we were not doing anything illegal. Given that this guy had a gun and uniform, and given the power of his position, I was rightly concerned.

And of course, the border guards didn't change at 2 am as Geoff's mate had reckoned, did they? The same guards interrogated us when I drove back into Greece a few hours later.

They were not amused, and they were absolutely sure that we were hauling drugs or something illegal. Which I guess you couldn't blame them for. They went through everything in the van, cutting the mattress up and pulling everything to pieces, thinking we had something to hide. They brandished large automatic weapons in our faces, kept us apart and questioned us for several hours.

Fortunately, we both stuck to our original story: That he (the van's owner) had received a phone call after we had crossed the border into Yugoslavia saying that he had to return to England ASAP.

Just before they finally let us go, the big bloke came up to me and said, with his face inches from mine, 'You don't think I stupid,' and, with a bark to his minions, reluctantly let us through. We were very fortunate. They could have thrown us in jail and thrown away the key.

I could have been in the slammer, and I still don't think Dad would have taken my phone call. Although I'm not sure those boys would have stuck to the 'one phone call' rule.

After driving the van around Greece for a few months, I decided to purchase a four-cylinder Honda motorbike to travel around the rest of Europe. I needed to sell the van and remove it from my passport before I could leave Greece and travel to London to purchase the bike. I also needed to recoup some funds on my van for the bike purchase.

Somehow, I made contact with some unsavoury characters who could get the stamp removed from my passport. You know, the old 'friends of friends' bit. As I walked with a group of them down the back streets of Athens, supposedly to get a judge to remove the registration from my passport, it felt like this was getting serious. I had the uneasy feeling that I might have my passport stolen and perhaps worse.

Tony (Hully's older brother) was away so, as the only other person I knew in Athens, I desperately contacted Theo Christo for his help translating the rapid Greek they were speaking. After picking him up in my vehicle, I talked to a few of the group members. They told me to wait while they supposedly did the deed. Theo Christo, in the front passenger seat, had heard enough to know he wasn't satisfied with the proceedings.

'Ella, Greiig, ella (Come, Greg, come)', he said. I interpreted this to mean I should assist him in getting out of the vehicle and dealing with these young rogues face to blind face. I guided him down to the street level, and he confronted the small huddle in no uncertain terms.

The language was loud and powerful, and I watched with glee as their arrogant body language crumbled. They had thought this was going to be money for jam.

I understood a bit of what he said, as I had been working on the boats, so I got enough feedback from the cowed audience. They clearly understood they should not do the wrong thing by this young Aussie tourist, who was a relative of his – I liked that bit – and they should not disrespect him, Theo Christo, an older war veteran and a very disabled man.

I repeatedly witnessed this respect for elders and the disabled, such as blind people, in the Mediterranean countries. Even gangsters were highly dubious about crossing this cultural line. I certainly got the feeling they weren't going to cross the older man. Respect for elders is reinforced by considerable social pressure within the culture.

Theo Christo asked me for the money I was to give them – 2,000 drachmas, about 50 Aussie dollars in those days – and told them he would provide them with half of what they had asked me for, because they were ripping me off. He said that if somebody didn't get that passport fixed immediately, it would be less than that. Then he twisted the money in his fingers with considerable precision, considering his one hand, and handed a note to them.

One of the group rushed off with my passport. I was a little concerned seeing it disappear around the corner in his thrifty little hands and wondered if I would see it again, especially knowing passports were highly valued documents on the black market. But after Theo Christos' warning chat and seeing how the young group had capitulated to him, I felt more assured about its return. Anyhow, whether I liked it or not, I had to trust he could get the job done.

Theo Christo and I leant back on the van for support, and the sheen on his plastic face seemed to form a wry smile as we waited for the kid to return. The old man said quietly in his few words of English, 'It's OK, Greiig, it's OK', and asked me to light his cigarette.

Although it was impossible, even as a card player, to accurately read his rebuilt face behind the skin grafts, I could see that he was satisfied with the situation. And, sure enough, the kid soon returned with the signed-off passport.

At last, I was free to travel to London and buy my bike.

I will never forget Theo Christo, a man who gave nearly everything to his country. His confidence as a blind, one-armed man and how he tackled life gave me confidence as a healthy young bloke. If he could do it, I could certainly take on anything I faced.

It was probably my first real encounter with a person with a considerable and severe disability, but there was never any thought of taking a back step, or any thought of retreat with him. He would hook onto your arm with his stump and say in his rough old gravelly voice, 'Ella, Greiig', and you felt like you could go out and take on the world as he did.

A great man. And it was a pleasure to have made his acquaintance.

RIP, Theo Christo.

The next thing I knew, I was travelling to London and Gatwick Airport. I was a little concerned about whether everything would be OK, given the circumstances. So I was anxious as I waited for the English Customs officials to stamp my passport. If there turned out to be a problem, saying, 'Oh, sorry, I did a deal with some gangster kids in Athens to get a judge to sign the vehicle off my passport' to the passport officer probably wouldn't have passed the test!

Flying to London on the concession flight I was on, it seemed that Customs had word of something in the wind, because they gave us the most incredible security checks and searches I have ever had. Every other passenger, it seemed, was taken aside to the frisk room. But the gangster kids' work had done the trick, and there were no passport problems there or elsewhere in Europe.

The Mercedes and Pete the Dealer

Taking a step back to my time in Athens, some of the most amazing experiences I had were with Tony whenever he asked his dad if we could borrow the Mercedes to go out on a Friday or Saturday night. Tony liked to take me to the Athens suburbs of Glyfada and Plaka (he was Greek Australian and spoke Greek fluently, as did Hully), where there were many bars and drinking spots, partly because of the US Air Force base nearby. Of course, where there are young men in uniform with money, there will be young girls aplenty.

The professional girls, if not perched in bars waiting for you to buy them a drink, were on street corners. These weren't seedy little dark alleys; these were corners on the main street. They had the eyesight of a startled emu when it came to seeing a nice new Mercedes approaching. The population of girls would swell in number from a handful to ten or fifteen per corner, with the additional girls appearing from nowhere when a flash vehicle was in their sights. It was an education for a young bloke, seeing what the potential of money could do.

Not only did the girls appear quickly, but they could also disappear just as quickly. When that happened, sure enough, if you looked back behind your vehicle, there was often a police car coming into view.

The owner of Angie's Bar in Glyfada (pronounced Glifather), Mick, was a mate of Tony's. Mick was a hell of a character. He loved Indian arm wrestling; it was a popular event in the bar, especially with all the US Air Force personnel as regular customers.

Often Tony would say to me, 'Go on, have a go', but for ages, I avoided it. But one early morning, after a few drinks, I got involved, and it got down to Mick, the owner, against me. Although not as tall as me, he was a thickset man and had been regularly arm wrestling in his bar for years.

Anyway, to cut a long story short, I managed to win. He was an excellent sport about it and bought a round for the bar. Then he asked me if I could do some security work for him: 'Just keep an eye on things', he said.

What he really meant, I was to find out later, was to keep an eye on his girls – his working girls. The girls were often not respected by their customers. It was something I could not understand and will not forget. As nice as Mick was, I didn't work for him for long, but it was one hell of an experience and I think puts life in some perspective.

While talking about Mick and his bar, I must tell you this story about a fellow who was a regular customer. He worked around Athens, and I reckon he was the smoothest character I have ever had the good fortune to meet. Because he had such a flexible definition of what was legal and what wasn't, let's just call him 'Pete'. His definition of right or wrong and black and white was sometimes inclined to get mixed up, but you wouldn't call him beige. Pete was anything but boring.

Pete was what you called a 'wheeler-dealer' and could get anything you wanted, from hot women to a hot car and everything in between. What amazed me most about him was that he was just such a nice guy. He was a good-looking bloke with a great personality; everybody liked his company. He had all the tools of the ultimate friendly crook and didn't seem to have a nasty bone in his body. Of course, there were always gorgeous girls around him. And that gets any young man or woman's attention – how does he do that? What's his trick?

Now and then, you would see Pete wave as he drove by in the flash vehicle that he was looking after for a lady he was having a fling with at the time. Her wealthy husband was flying in to meet her in a week or so from an overseas conference, and Pete was just being a gentleman and showing her around.

I didn't spend as much time as I probably would have liked to with Pete, but I ran into him at a bar or around town from time to time, delivering something for somebody, a particular car part, an introduction to a lady, or whatever; there was always something going down with Pete.

I got to thinking, even if you were doing something illegal with Pete, he had this way about him that convinced you that it wasn't problem at all. It seemed like life was just one big joke to him, and it was infectious.

Now by complete fluke, and unbeknownst to me, when I took that charter flight to London from Athens, who should be on that same flight but Pete and his gorgeous girlfriend; let's call her Jane. I was surprised they were not sitting together on the flight. Oh Greg, you were so naïve at twenty. But I was thrilled to see them, as I hadn't managed to say goodbye to them with all the drama going on with my van sale.

I know what you're thinking, and you're right; why hadn't I got Pete to assist me with the van sale and passport? But unfortunately, I just hadn't seen him for a month or so. No doubt tied up with some monkey business somewhere.

We were going to Gatwick on this cheaper charter flight, and after a quick chat, we agreed to catch up at the airport for a drink. I had been on a few flights by that stage and had a fair idea of what to expect. Little did I realise that this experience would be considerably different. Now it didn't even occur to me that because it was a 'cheaper charter flight' from Athens to London, it might have been of interest to airport security. Perhaps the powers that be had got a whiff of something or maybe, dare I say it, the mere fact that Pete was on board might have raised suspicions. They went through every one of us with a fine-tooth comb at Gatwick Airport. I mean every other person – into the little room for a closer examination, smear tests of deodorant cans, going through all your toilet bags – the works.

Now while all these checks and thorough investigations were going on, who should rock up in line not far from me but Pete? I must say, as much as I liked him as a person, I was pleased he was further back from me in the line. We were all aware of the stories of somebody being 'loaded' at the customs counter by a passer-by. And, frankly, nothing would have surprised me with Pete.

His line was just over from mine and was travelling a little quicker, allowing me to see and hear Pete in full flight mode. At his best in charm. He had been allocated to one of the few female customs officers. I don't think even Pete, with all his skills, could have arranged that. Everybody was taking these security checks very seriously, and some were getting pretty frustrated with the extra attention and time taken – all bar Pete! He never batted an eyelid.

When he got to the counter with the female attendant, he started to unbutton his shirt to the waist and said, 'Would you like to have a look at me, darlin'?' The girl, in a very unamused fashion, said, 'No sir. That's fine, thank you. Just let me have a look at your overnight bag.'

There were still a couple in front of me in my line, and I could see Pete's every move as he unzipped his shower bag and opened it up for her to rake through all his condoms, shaving cream, men's aftershave and other contents. After she inspected all the products within the bag (and I'm sure Pete, with his humour, had intentionally left the condoms open for this public gathering of security staff and passengers), the young lady finally waved Pete on without further comment.

Peter thanked her for the pleasant experience in his cheerful way, and we both progressed towards the train, waiting for Jane to catch up with us.

After purchasing tickets for the train to London, Pete began to laugh raucously, and when Jane joined in the laughter, I could see there was more to this personal joke than I was aware of. Pete, who could read people like a book, of course, another of his essential skill sets, nodded for me to come over and sit next to Jane and himself on the train. While confirming nobody else was close enough to see, he pulled his stomach in and his belt out. I got a quick glimpse of a large quantity of what I guessed was marijuana, partially wrapped up in newspaper and what seemed like greaseproof paper, possibly to avoid the sniffer dogs. I couldn't be sure, but he may have had a kilogram or more on him, and it doesn't matter anyway. It was the amount of hide he had that amazed me. Remember that, in 1976 in the UK, one could do considerable time in prison for that amount of drugs.

He smiled at me, not out of arrogance or being a smart-arse, but as if it was just part of his adrenaline-packed life that he enjoyed so much. He was intoxicating to be around, and he was everybody's friend.

That was the last time I saw Pete and his gorgeous girl, as we went our different ways in London from there. We were meant to meet up, and I was so pleased to have made his acquaintance and friendship over many months in Greece, but as often happens, when you're travelling, you go your own ways.

I missed not meeting Pete and his mischievous grin again, not that I was interested in any of his illegal pursuits, though I'm sure I could have gotten involved if I wanted to. But I missed his cheerful company and his silky-smooth skills and personality, one I think many young men would have aspired to. What a great Hollywood blockbuster Pete would have been.

I've often wondered what Pete would have been like as a business partner. Indeed, it would have been an exciting career move – never boring.

My biggest frustration – I didn't see his lovely lady friend again.

Working on the Charter Boats in the Mediterranean

I remember when I first walked up the pier in Piraeus, the port of Athens, and saw the 'squillion'-dollar yachts moored there. Many of them had a backpacker on board, acting as a caretaker. They used to clean the decks, do a few odd jobs and start the motors occasionally in return for pretty much free accommodations. Owners worldwide would fly in with family and friends and take these luxurious toys out to the Greek Islands for a few weeks for business, pleasure and whatever.

Michael (Hully) introduced me to Peter Boulton, who had a motor charter boat, the *Bandersnatch.* Peter needed some aft cabins built, as well as antifouling of the boat's underside. As a bush kid born in the desert, this task was entirely out of my range of experience, but I explained to Peter I could use some common sense to build the aft cabins. I had done a bit of carpentry and would give it a go.

Two months later, after completing some work on board, we left Piraeus, heading for the island of Salamis. By this time, an American guy named Pat Collins had joined us. Apart from liking peanut butter and jam together on toast, Pat was a good man. Our dramas only really began when we got to Salamis.

When we got there, we slipped the boat up on the shore. The deck would have been about 6 metres above the ground, so we could see what was happening everywhere. We used to go ashore to buy groceries and experience the local scene, but mostly worked all day on the boat.

One evening, I went into town with Pat, Peter and Peter's English girlfriend, Wenda Dickens. In the main street, it was pretty apparent that all the eligible young men were on one side of the road, some in small groups and others by themselves, and on the other side of the street were all the eligible young girls, being escorted by their parents or family members. Although it was 1976, it seemed to me like some ancient cultural custom playing out.

We were among it before we realised it was all happening, and we were gobsmacked. Though, in a historical way, I thought it was quite beautiful, as everybody was upfront about what they were doing: These girls are available; let's see the men and vice versa. You couldn't say it wasn't out in the open. It was like watching a play, but it was real life. An amazing experience.

But I must offer a proviso here. We were tourist visitors to the island who didn't speak much Greek and didn't have much interaction with the local inhabitants. Although it looked very legitimate, it may have been some production by the local community. However, it was certainly compelling and entertaining

There was another issue that we hadn't planned for on this beautiful island. You see, Pat met this lovely Greek girl and they were very keen on each other. She had come on board once or twice while the boat was slipped. When her big brother and dad found out we had completed our work and were going to ease the boat back into the water, leaving the island, all hell broke loose. We were to learn, in our ignorant way, that the relationship was OK, but Pat needed to take this relationship to fruition and marry the girl. That's where the cultures seemed to differ markedly.

We could see this was going to become a serious issue and hid Pat underside for the last thirty-six hours before our departure. More by good fortune than good planning, we had a single plank walk from the shore up to the boat deck, which could easily be removed to stop anybody boarding, much to Pat's relief. Big brother and the old man were at the end of the gangplank with a machete in hand as we left. We had interfered with the island's customs or something.

Now that the aft cabins were built and the boat had been antifouled and painted, Peter didn't need any more assistance from us and was ready to take his next charter.

Our last evening on the *Bandersnatch* was exceptional. We went all night, Peter and me bouncing off each other, telling yarns. As an English charter boat skipper, Peter was in his element, telling tales after dark.

One of them, I will never forget.

Peter had spoken several times about some Arab customers he'd had since starting his charter business. They were wealthy men in the oil business then, but when he first met them, they were battling financially and admitted that they'd earned their living as pickpockets as younger men to get their start.

He told us of an experience he'd had with them on a similar evening to the one we were having, with a few drinks and lots of laughs. They'd mentioned to him that they were still making their living picking pockets. Peter, being a straight shooter, suggested that was bullshit. So, sitting across the table from him, they explained that he wouldn't know if they had stolen his watch or his wallet. Peter again said bullshit – they wouldn't be able to get his watch off without him knowing, even after a few drinks. They casually said OK and just went on with the conversation.

After a while, one of them asked Peter what the time was, as he was tired and might go to bed. Peter went to look at his wristwatch – it was gone. His charter clients proceeded to give him back his watch, his wallet, his pen, and some coins he'd had in his pockets. Peter was dumbfounded and utterly amazed at what they had done.

Now he was on guard, and he carefully watched their every move. He felt comfortable that they had not done him again, but when one of them left for bed, he very carefully placed Peter's wallet and valuables back on the table with a knowing smile.

As Peter said, they had pickpocketed their way into a legitimate oil business. Who says crime doesn't pay?

We both agreed there and then that if you were unfortunate enough to run into a good pickpocket anywhere, any time, you wouldn't even know you were done until it was too late.

Our relationship sadly concluded that night. As Peter and Wenda motored away from the main jetty in Piraeus, I thought what a lovely few weeks we had all had. I was eager to see where my travels would take me next.

During the last week before the *Bandersnatch* left port, a South African millionaire named George Rouge had sailed across from Durban and tied up close to us. He'd invited me aboard a few times for drinks, as happens with boats and yachts along the quay. One time, he asked if I would be interested in finding a few girls who would like to travel around the Mediterranean for a few weeks on his yacht. I mean, it was a tough decision, but...

We would sail about and pull up at any of the islands we wished, drop anchor, and get restaurants to serve us on board with all the eats and drinks we would like, all expenses paid for everyone, of course.

Well, that's precisely what happened. It's the sort of stuff you dream about, isn't it? But I had to find the right type of girls first, because weeks at sea with the wrong kind of people could ruin the whole show.

So after confirming a few more facts and details with George, I set off for Syntagma Square, where all the tourists were, because George and I felt it was more sensible to find English-speaking tourists than Greek girls.

Sitting in Syntagma Square having a Turkish coffee, I could not help but overhear two English-speaking girls who were having trouble getting their money out of a Greek bank. They sounded desperate. I introduced myself and said that perhaps I could help because, after living with Greek families and working on the boats around Piraeus, I had picked up a fair amount of Greek – enough to get by, anyway.

One of the girls, Ellen, was from Canada, and her friend, Georgie, was from Washington, DC. It seemed that Ellen's parents had sent some money over (or wired it, as they used to say), and the bank was just not that interested in handing it over.

I'm sure being a bloke helped, and slightly better communication was enough to get the deal done; they were so pleased to get their money.

After that win, I figured I at least had a chance of taking the next step, so I invited them back to the square for coffee and said I had an exciting offer for them to consider.

'Yeah, right', they said, 'and there's no catch!'

But after explaining the situation to them, they agreed to meet George and said they'd make their mind up then. I'm not sure I would advise my daughters to consider such a decision in 2022, but it confirms the saying, 'faint heart never wins fair lady.'

After inspecting the yacht, the girls were convinced of George's authenticity, though it may also have been down to the lovely champagne he had purchased. Either way, before long, we were sailing out of Piraeus en route to Cape Sounion, where we anchored and commenced an incredible few weeks' cruising around the Mediterranean.

In Kalamaria, a pretty little town nestled in the hills, we had a lovely fish meal, listened to bouzouki music and drank *domestica* wine. At Chalkis, on Euboea, we fluked the old bridge of Evripos being open, as apparently it was often closed and used for land traffic, meaning boats with high masts couldn't gain access. After a couple of rainy days in the coastal city of Volos, we set sail for the beautiful island of Skiathos. It was my twentieth birthday, near the end of May. We celebrated on board with a dinner of lamb chops and veggies, and the girls gave me a miniature 750 cc motorbike because I had mentioned I wanted to ride around Europe on a motorcycle. We sailed around many other beautiful islands and dropped anchor into lovely clear shallow water where you could see pristine marine life. It was paradise.

All good things must come to an end, it seems, and eventually, we sailed back to Piraeus, said goodbye to George and thanked him for his kind hospitality. The girls stayed with me for another week, looking around Athens and the surroundings, as they had not been in the city as long as I had.

When it came time for them to go home, I was very sad to see them go; I remember Ellen, who played the guitar, playing the song 'Leaving on a Jet Plane' before I took her to the airport. I made the near-fatal mistake of paying the taxi driver nearly fifty bucks Aussie to get us to the plane because we were late, and in those times cheaper flights were not transferable.

We spent half the time with two wheels on the pavement at the lights and elsewhere, overtaking cars madly and it was a heart-in-the-mouth trip but effective, because we arrived just in time for Ellen's flight. I was very sad to see her go.

It broke us all up, but we have kept in touch, and Ellen and Georgie have been to Australia a few times since then. They came to Curbur and stayed with me for a while. Ellen was very fond of my mum and vice versa, with both hugging and crying when she left to return to Canada after her visit. It was the last time they would see each other before my mum passed away.

They were great supporters of my writing my first self-published book in 2016. And it all started in 1976 in a public square in an ancient city on the other side of the world when a simple favour turned into a lifelong friendship.

CHAPTER 8

'Ya Can't Miss It, Mate.'

We all know what it's like to get directions from somebody who tells us everything except what we need to know, and when the first thing that they say is, 'You can't miss it', you know you're shot before you start.

When I was around twenty-one or twenty-two years old, there had been many years of droughts in the Murchison, and our family leased a farm at Newdegate to put the last of our merino ewes on so they could have progeny for the future.

The first time I travelled directly across from the Murchison to Newdegate, in as straight a line as possible, I took some isolated routes on gravel roads from the north of Newdegate. Some 930 kilometres and eleven hours of non-stop travel.

Now, there are many lovely people around Newdegate, but I was to find there were also some cousins married to cousins, and maybe some Deliverance-style banjo playing out in the back blocks. It was pretty isolated, even by station standards.

After crossing the fourth gravel crossroads without any signage and, as none of these roads was on my map book (no GPS back then), I decided to call in at the next farmhouse I saw that was close to the road and get directions.

Eventually, a farmhouse came into view, and I thought I would try my luck, not even knowing if anyone would be home.

I pulled around the back on the main worn track, because that was the entrance, of course, and the first thing I noticed, hanging from the roof cables, was an old, skinned roo (kangaroo) carcass that had obviously been there long enough for most of the flies to lose interest.

I was yelling out to ask if anybody was home and contemplating whether I would get past the roo, which I was dreading even from this distance, to the partly busted back door.

A barefoot bloke with just shorts on came out to the noise of dogs barking, scratching his crotch. In my young, energetic, fertile brain, I wondered if he had just come off a night shift, was just late in rising around 11 am, or was heading for a midday siesta. By the looks, I think it may have been the latter.

'How ya goin', mate?' was his first cheery comment, in quite a happy voice, compared to his looks and surroundings.

Probably not many visitors came this way, was my guess. I apologised for disturbing him and asked him for directions into Newdegate.

'Yeah, naa worries, mate, ya can't miss it,' he replied, still scratching something that obviously had a terrible itch, as by now he needed his complete hand. A few fingers hadn't done the job.

I was grateful for the kindness, but with the combination of roo stink and severe scratching, I retreated a couple of steps back into the fresh air and the sunshine, confident that I would be able to hear directions from a little further away.

'Murchison Plates...? Ah mate, a long way from home, ah?' he said, surveying my V8 Sandman Ute thoroughly, noting the dog in the cab, the drum of fuel in the rear and the motorbike tied on the back.

'Yes,' I said. 'We've leased a place just north of Newdegate for six months to agist some sheep on, and I'm going there after picking up some stores in Newdegate.'

'No worries mate, yeah, ya can't miss it,' he said, quite seriously. 'Ya go out here, mate, out my front gate, the wheat crop there on ya right's real good this year, go through the gate and out onto the main road.'

'Yeah, that's the way I came in. I've got that,' I said.

'Yeah, that's it,' he said. 'Ya go out the gate and turn to ya right and go down about 3 miles [we had only gone metric thirty years before], and ya come to a bridge and cross a bit of a gully there. You'll know it when ya get to it because all the rails are painted real white, and there's a big batch of trees on the right-hand side, it's pretty narrow, and only one vehicle can cross at a time. My grandfather built that bridge in 1945, just after the war, and all the timbers underneath came from Pemberton, but ya probably won't see those timbers when ya go over.

'And on the left-hand side when ya come off the bridge on the other side, there's a big snakewood tree 'bout 100 years old; my father planted that for me grandfather when he was building the bridge, and there's a little plaque there just past that snakewood on the left-hand side. The white rails are real easy to see, ya can't miss em.

'And it's a little rough on top, so go a bit slow, not too quick, ah? And there's usually a bit of water there but pretty much dry now.'

He grabbed a good hold of his pants again in an attempt to get that itch and said, 'But anyhow, you don't need to go as far as the bridge. The turn-off you need to take IS BEFORE THE BRIDGE, 'bout half a mile this side... Ya don't need to go far as the bridge.'

Oh. My. God.

CHAPTER 9

Mosbar Gooden

Content warning:
Animals are harmed in this story.

What's this thing we do as humans where we add a name or two to our loved ones, be they animals, partners, grandparents or whatever? It just seems to be a silly little thing we do as people.

Mosbar should have been just 'Mosbar' but ended up being 'Mosbar Gooden' because he just did. You know, like lady girl, Jane dear, Janet darling or Nanna Ya Ya. He became 'Mosbar Gooden' and, although he was a male dog, he had a beautiful temperament and wouldn't hurt a fly.

He was a purebred, short-haired, border collie pup that I purchased and imported from Wales via quarantine in 1976 when, at the age of twenty, I was travelling through Europe on my four-cylinder Honda. Both his male grandparents had been champions at the Cardiff Royal Show in 1967–68 and the experience of witnessing the working sheepdog trials way up in the Cambrian Mountains was amazing. It made me understand just how good these dogs were and what they could do with a mob of sheep. These scrubber mountain sheep, built like tanks, used to run up and down the narrow little passes on the mountain cliffs, just wide enough for an animal with very sure footing to travel without falling. These dogs were unbelievably good at keeping them together in treacherous terrain and mustering them down to the paddock far below – amazing to watch. Imagine these working dogs circling back in a wide arc so as to be in a head-on position with any escaping sheep, blocking their exit gently but persistently. They would have been marked down considerably if they had traversed the same rocky pass as the escaping sheep and, perhaps worse, split them up.

These dogs could muster a blowfly into a bottle, I'm sure. It was like nothing I had ever seen in Australia, although I'm sure it was happening somewhere there. Time and time again, they worked to perfection. The breeding had a lot to do with it, hence my purchase from these well-bred dog lines.

I hoped to see myself replicating those performances on my return home to the station in outback WA, which also had its challenges for a working dog. Even for an experienced stockman in the mid to late 70s, a dog holding a mob of sheep or goats together in any country without losing a single animal – every time, without fail – was quite a fantastic accomplishment. And let me tell you, wild feral goats being worked for the first time will test any dog out.

The dogs had to be wide-working (staying far enough away to not pressure the livestock unduly) with plenty of 'eye' (a genetic trait bred into working dogs connected to their complete and absolute concentration). Directly affected by movement, if the sheep makes a move or turns to escape, the dog will immediately adjust its position to block it. It's all about keeping the mob together. They also had to go all day and to be as hard as a nail.

Apparently, Mosbar had been the runt of the litter. I found this out later; it's not necessarily something you tell the interested buyer. However, I know some dog owners who ask for the runt of the litter. He was a very gentle, sensitive, little dog, relatively light in the frame under his slightly hairy coat, with three of his four paws having white socks.

You have experiences with dogs in the bush which are inclined to bond you to them for their life, in appreciation mostly, as I did on this occasion with Mosbar.

I was goat mustering up at Doorawarrah Station, in the Gascoyne District east of Carnarvon WA, along the Gascoyne River one day, and I had two or three working dogs with me, including Mosbar Gooden. I was up there because, after years of droughts at home in the late 70s, I felt disposed to arrange with some other pastoralists to muster goats on their properties for a cut of the sale proceeds.

The sandflies are particularly bad after a storm, which split the feral goats into small, scattered mobs, all keeping apart to minimise the number of the pesky blood-sucking predators in their face. Separating, in turn, made them agitated and not inclined to mob up, making them more challenging to muster together for collection and yarding.

I decided that as my motorbike noise was driving the already separated mobs further apart and achieving the complete reverse of what was wanted, I would leave my bike and run on foot with the working dogs. Then we would sneak onto the next unsuspecting mob of goats before they knew we were there and continue towards the next mob for collection, mobbing up as we went. My dogs rarely barked unless under real pressure.

However, as we were running into more little mobs of scattered goats, the lead of the mob was getting further and further away in front of me, where Mosbar was continuing to block up not only the lead but fresh goats we were coming across on our journey. This was in sandhill country. Every time I found myself gasping for breath after running kilometres keeping the tail up with two of the dogs, I could see Mosbar getting further and further away in the distance, trying to block the advance of fresh and fit wild feral goats and effectively the lead of my expanding mob.

At one stage, I could see Mosbar way out of earshot, atop a large sandhill, blocking some giant billy goats standing high up on their hind feet as they do in attack mode to scare a would-be predator, as he was retarding their progress to run further away.

As I panted from the base of one sandhill to the next, having travelled some 8 or 10 kilometres back and forth on this tail of goats which were now slowing, I could only occasionally see my little friend. I knew he couldn't block up fresh goats with too much pressure and had to backpedal occasionally to let them think they had the upper hand before settling them down back into the mob. Hence the need for my long-running journey. Patient Mosbar had not been too forceful initially, which would have forced the goats to break past him and escape away into the distance. This way, they were mobbed up with their mates, and the predator was far enough away for their comfort.

Making the summit of a large sandhill near the Gascoyne River, I noted he had finally been able to block the lead goats back enough that he was only 500-odd metres ahead of me. And I could guess I probably had 400 or 500 goats mobbed up from lead to tail.

It wasn't just the potential financial gain of perhaps twenty dollars per head, average, but the fact that my little Mosbar Gooden had won the day against a formidable challenge and allowed us all to have a win. We had a long way to go before we yarded most of them, but it was a very memorable day for the little man, who was confident but not too complacent, as he just came up for a little pat after yarding up with all dogs as if to say, 'Did you see how I blocked that lead up, Dad?' That day, his grandparents' genes from the Cardiff Royal Show came out to play, and it started a significant relationship between us. He was only around two and a half years old at the time, but I knew I could expect great things from him.

Mossbar was now about five years old and, after having years of dry seasons on the station, my father and I arranged to place or agist the last of our breeding ewes on unharvested crops in and around Mullewa. Many of the farmers had not even bothered putting their harvesters into the paddocks because of the limited growth of the crop but these areas were useful for sheep grazing, and few of the farmers ran sheep anymore. Still, they could gain some lesser funds from offering agistment of livestock on their properties to help make up for the crop shortfall.

The use of cropping country around Mullewa was fine whilst there was still grain in the heads and still feed, but when that finished and dried off, we needed, ideally, some greener pasture to mate these ewes down on and I managed to locate a property on the Moore River near Guilderton. A farmer there wasn't using the country on his farm and wanted to agist sheep and they could graze right down to the river to drink, which, because it was always full, also acted as a river fence, unlike much of the country these sheep were used to. I had already trucked larger numbers of sheep down there previously and decided to take the rest of the Mullewa sheep down there as well, when we ran into a problem.

I had visited the property at Guilderton a few times and could see the feed was running low in a few paddocks. In discussion with the farmer, we had agreed that he would repair a couple of gates in the nearby paddocks which were not stocked and move my sheep into them.

I made a few phone calls to him but he still hadn't managed to fix the gates for some reason and my last inspection, a few weeks earlier, indicated that the paddocks the sheep were in would now be getting very low, so I was starting to become concerned about the sheep's condition.

He was getting paid per head of sheep grazing on his property, so I felt that surely it was in his interests to care for his investment. But I was losing confidence by this time and although I still had sheep at home on Curbur Station to care for as well as usual jobs, I said to him I would be down the next week, a 600-kilometre trip.

On my arrival at the farm, I had my heart broken, as the gates still hadn't been repaired and some of the sheep were barely able to stand, they were so weak from lack of feed. It was only grass and pasture in this farming country, no grain from crop stubble, and once the grass had gone there was not an edible tree or shrub in sight for these pastoral sheep.

But worse was to come, because when I jumped on the motorbike to properly survey the situation, I saw to my dismay that when some of the weaker sheep had gone down the steep riverbank for water, some had collapsed and drowned in the river.

The picture got worse, but I won't describe further, because it won't bring them back or change what had happened.

As a young man, I was stunned as to how anybody could allow this to happen. Was it lack of interest, lack of effort, laziness, what? But on maturity there are often reasons – some of them reasonable, some of them very poor. This was no doubt on the side of the latter.

Without asking or discussion I ripped open the gate in the corner into the next paddock they should have been in already and gently herded my sheep into it. Once the hungry lead had started, the others didn't need any more enticing to follow. I fixed the gates in the other corners in thirty minutes with my fencing pliers on the bike, while the sheep were still munching on the new pastures that didn't even have much growth, but were miles ahead of where they had been.

When I left the Guilderton location that evening, I don't think a ewe had lifted her head from the pasture since I had arrived that morning. In a week, when they were stronger, they were removed from that farm and the agistment cheque was way, way smaller than it could have been, and we removed many fewer sheep than we had brought.

But what was carved into my brain from that experience, among other things, was that people who agisted livestock on their property might have a dubious reason for doing so. I mean, what's the reason you're not running your own livestock on your property?

I understood the sheep and what they were about in this situation – they were only trying to survive. The problem I had was with the humans. So, I came back to the sheep we had at Mullewa, some 350 kilometres back north that needed to be trucked off that farm with that drowning situation spinning in my head.

On this day, when I arrived back at the Mullewa Farm sheep yards in the semi after dropping a load of our sheep at Regan's Ford, I found little Mosbar Gooden was lying on his side near my motorbike.

He was hardly moving; as I went to pick him up, I realised he was peeing blood. The slightest little tail movement was the only acknowledgement he gave me. I was trying to figure out what had happened. I could only think that the poor little fellow had been cornered in the sheep yards by a ram or something, because I knew he wasn't a great jumper, to escape.

The sheep yards were made of rolled-out flat metal from 44- or 200-litre fuel drums sitting vertically on the ground and joined end to end. Drawn into the sheep yards by seeing others loading or handling sheep, the dog would be eager to become involved. But in such a case, if a gate were shut, it would have been impossible for him to get out of the yard without assistance.

Stroking him gently, I was frantically trying to figure out what could have happened for him to be in this state.

After not thinking the worst initially, my second thought was that he might have got a snake bite. But that didn't account for him urinating blood or the way that, as I patted him, his whole body seemed to be in so much pain.

Surveying the scene, I started to suspect that Mosbar had been attacked by somebody on the farm, and perhaps they assumed incorrectly that it was Bondy's manager's dog. He had just arrived recently and was working for Alan Bond in reprocessing the farm. Bondy, a well-known WA business tycoon, notable for his part in the WA Inc fiasco in WA.

The farmer who had previously had this farm was recognised around the area. I think some people felt it was unreasonable for Bond to expect him to make farm purchase payments when it wasn't worth putting the header in the paddocks, due to such limited rainfall.

It just happened that I and a few others were agisting sheep on that property for payment when it was repossessed. It seemed Bond had the legal right to reclaim the property, and we were just in the wrong place with our livestock at the wrong time and had to remove them.

It took me some time and a few discussions on-site with Bondy's manager and others before arriving at my disappointing conclusion. Mosbar had indeed somehow been the victim of this local feud. What we saw later at the vets, on Mosbar's body, I don't think could be called an accident.

I'm just so sorry Mosbar was impacted by his involvement, as little as it was. And, of course, none of this local feud is any excuse for a perpetrator or perpetrators.

There was no vet in Mullewa in those days, and the only man I could find to look at Mosbar was a Mullewa local who did some minimal part-time veterinary nursing assistance to animals in town – a bit of stitching or some ointment for mange or whatever.

Little Mosbar Gooden was barely alive when I drove away, hoping for the older man to perform a miracle, but my gut told me the worst. When I returned the following day, he told me that Mosbar had died not long after I left the previous evening. But he had opened him up to see if he could find a cause of death and peeled the skin back to show me the black and blue bruises all over his body. I had to presume he had been kicked to death. I don't believe anything else would cause that type of bruising and trauma all over his body. I racked my brain to think of other ways such damage could be afflicted on an animal but could not come up with anything remotely feasible. If there were a horse on the farm, which there wasn't, one or two kicks could have been understandable, but his body was nearly completely covered.

I felt a sharp pang of pain, remorse and guilt go through me on witnessing that and wished I had not decided to leave him alone near my bike. I thought I was doing him a favour, not tying him up. That's what he liked best: to stay untied near my bike.

The poor little feller had tried so hard for me over the years, and it ended like that. Just heart-breaking. As happens after these experiences, you always wonder how you could have done things differently. I can appreciate readers thinking, 'Well, why didn't you tie him up or take him with you?'

Well, the truck was a hot old cab-over International, it was a long trip, and I knew he didn't like the noise and vibration. And if I had tied him up, there was always a chance he could get hooked up or knock his water over. He used to be very happy just staying with the bike, that way he could run free as he always had. And I didn't expect anybody else to be around, or even know that anybody else had sheep agisted there on the farm at that stage. I also wasn't aware until afterwards of the level of angst in town about the prominent entrepreneur repossessing the farm from a local after years of drought and, consequently, poor crops and limited income.

I went back to the farm to bury him; tears flooded my eyes when I saw the little hole under my motorbike that he had dug to camp in during my absence, as he liked to do.

Despite my many enquiries and frustration, I never found any answers as to who had been responsible for Mosbar's death. There was a closing of ranks. Nobody seemed to know what had happened. Years later, I spoke to someone in that circle at a funeral; he acted as if he was completely surprised by my story and said he knew nothing about the incident.

I guess you wouldn't, would you?

RIP, Little Mosbar Gooden.

CHAPTER 10

The Old Bull – Just One Step at a Time

I have learnt many lessons from the bush over the years.

Nature is perfect, although it is often a harsh and demanding teacher. What we see in the bush often gives clues about how to live our own lives sustainably. We should follow its ancient natural laws.

One such lesson came from a meeting with an old character when I was out doing a mill run one day. A mill run, or bore run, as some call it, is when you check and service the watering points on a vast property so that livestock can drink fresh, clean water. In this case, it was checking cattle on Wandagee Station in the Gascoyne region of WA, off the Manilya River, around 30 kilometres east of the Manilya Roadhouse, on the Northwest Coastal Highway.

Before I go on, I should try to explain to those who have not had much to do with the bush just how vast and isolated the land spaces are. For some perspective, my old home property of Curbur Station, in the Murchison Shire, was around 200,000 hectares (over 2,000 square kilometres), and Curbur is not a large station property, but probably mid-size. If you were to go up in an aircraft to 1000 metres on a clear day, near the centre of the property, you could look in any direction and all the land you could see would be part of the Curbur Station property lease.

Or, to put it another way, if you arranged for one of the fastest motorbike riders around and asked them to ride as quickly as possible around the Curbur lease boundary, it would not be easy for them to complete the task in a day. The time taken to travel such distances are still hard for me to fathom, and I was born and bred to it. Still, perhaps somebody would like to take up that challenge one day? And maybe raise funds for a charity? What a great idea for readers to consider.

But this is not about stupid bragging rights or how big your property is. It is about what the property produces in terms of heads of livestock. Because annual average rainfall is only around 200 millimetres in this part of the world and no crops are grown, the area of natural rangeland needs to be significant to sustain reasonable numbers of livestock. No crops are grown on stations, with the odd exception. But they naturally graze livestock on it for beef, sheep, wool and a small amount of goat production, as well as perhaps some station tourism such as retreats or the like.

So hopefully, that puts some context into these stories that readers can grasp, because some may be thinking, 'Why didn't he do this or that to fix his challenges?' as if it was within walking distance to the back shed.

On this day, I pulled up alongside a weary five- or six-year-old micky bull (as shown in the picture below) who would usually have weighed around 650 kilos live weight. This bloke weighed much less in his present state. He was worn and battered from fighting and competing with other bulls to protect his patch. He seemed to have come off second-best several times.

I stopped my 4WD alongside him over the fence, only 2 or 3 metres away. He would have normally run away from me in such circumstances but was unable because of his disabilities and injuries. I sat and watched him treading heavily and slowly, just managing to put one leg in front of the other. He looked severely crippled and was probably on a journey towards a water trough at Mulga Well windmill, which was at least 5 kilometres down the cattle pad.

A cattle pad is a narrow track that the cattle (and other animals) use to travel to and from watering sites. A regularly used pad has often been chopped up into powder or softened 'bull dust' from years of cattle hooves working that soil. The pad winds its way around trees and fallen logs, via the most direct route, to and from the water trough. The greater the number of cattle using the pad, the deeper and dustier it is. As the size and wear on the stock pads are proportional to the number of livestock watering there, somebody with experience will be able to have an educated guess at livestock numbers watering at that waterpoint just by examining the cattle pads.

This picture was supplied by Alick Edwards

This old scrub bull had never seen the inside of a cattle yard in his life (because if he had, he would never have been released: he would have been trucked to market), so he must have been fighting with his fellow competitors for many years. He stopped and stared at me with his big dark eyes, unmoving and unthreatened, as if to say, 'What would you know?'

He continued to hobble along the cattle pad. He had missed a few musters because the stockman had been unable to wait for him; he would not have been able to travel fast enough to keep up with the main mob to the yards.

His head seemed huge compared to his bruised and battered old body, as is the case when the damaged muscles and nerves in the body wither. His big eyes surveyed me as he snorted snot in my direction in disgust. I was delaying him from his long and arduous task when he could barely put one foot in front of the other. The thought of him having to travel another five kilometres into Mulga Well, that life-giving water trough, on this sweltering mid-January morning, concerned me. Most of the other cattle in the sizeable 50 square kilometre paddock would have travelled to the water earlier, in the cool morning, but his progress was so slow and arduous. In the ten minutes I watched him, he had only walked about 100 metres. 'Gee,' I thought. There would be little chance, given the heat of the day, to make the distance before sundown in his condition.

He hobbled on in his emaciated frame, and although he looked partially defeated, he seemed genuinely determined. I admired his extraordinary persistence in achieving his goal and considered how strong that survival instinct is in all of us living beings.

As I drove away, leaving him in the dust behind me, I could not stop wondering if he would make it or not. Nothing would stop him from trying his best to get there, just putting one foot in front of the other. I could not help but reflect on our shared moment and his intense gaze.

I arrived at my first job some 10 kilometres away to find that the cattle had broken the ball tap off at Warri bore, and precious, lifesaving water was spewing onto the dry ground, much to the disgust of the bellowing cattle. The job required tools and parts that I did not have with me in the vehicle; all I could do at the time was turn the valve off at the tank to at least save the remaining water.

I had to drive to the homestead, 20 kilometres on the most direct route. After grabbing the necessary parts and tools from the workshop, I quickly raided the freezer for some more water and fruit. It would be a bloody long, hot day; I could feel it coming.

On my return to Warri bore, I was concerned about the potential outcomes of the unfolding chaos there. When cattle do not get their much-needed water, they can get restless and potentially forceful, pushing, breaking and smashing stuff. They can be terribly powerful when distressed; many watering facilities have been bent and twisted by this mayhem. Metal, steel pipe, it does not matter. It can all end up in a twisted mess. It was better to shut the water off entirely, as I had been able to do at the tank. Then, when the whole area was dry with no drips of water or smell of water to draw the cattle's attention, they might go easier on the infrastructure.

On my return to Warri bore, I saw the tank had made up a little of its lost water supply during my absence. I quickly repaired the ball tap and float (which lets the water run into the trough until it is full and then shuts off). Once I turned the outlet valve from the tank back on, I stood nearby for fifteen or twenty minutes, so I could watch until most of the cattle had drunk and calmed down a little. This gave me time to peel and eat that orange from the homestead fridge I had been looking forward to.

An old cow nibbled the orange peel I discarded onto the ground. I saw by her belly that she had already had her fill of water and was taking my orange peel for dessert. She reached up with her long tongue to lick my hand as if to say, 'That was alright, what else ya got?' Her tongue was like a wet rough rasp on the back of my hand, and I realised why chunks of grass didn't stand a chance.

With the watering repair job done, I could leave Warri and continue checking mill-runs and servicing other water sources on the western side of the property. There were still another dozen to be checked, so I had to keep going.

But because of the problem at Warri and the trip to the homestead, I was forced to take a different route than I'd intended – by complete coincidence, via the Mulga Well, where the old bull had been heading when we last met.

As I was driving there, I figured it must have been about four hours since I had seen the old bloke. Upon arrival at Mulga Well, I was astonished to see him at the trough already. There he was, head down, nose out like a snorkel, so he could still breathe without having to remove his mouth from the cool water.

The bull stopped drinking, lifted his head and turned towards me with a massive snort of water and snot through his nose. His belly was as bloated as if he had swallowed a 200-litre drum. I gazed in wonder and respect at the sheer willpower of the beast who had made it here after his long journey, especially given the timeframe and his disabilities.

He just gave me a disdainful look as if to say, 'You did not think I would make it, did you, you dimwit?' What determination and perseverance he must have possessed! I smiled and took a picture on my mobile to confirm that I was bearing witness to his persistence and drive. Having done so, I went away feeling inspired by the resilience of the old bull. I reflected on the lesson I had learnt from his determination and natural strength. And the life lesson may well be this: When you have a tough challenge to overcome, break it down into smaller steps, keep focused on your primary goal – and get ready with a look of disdain should anyone doubt you or try to impede your journey!

As I said at the beginning of this story, nature, animals and the environment all have a way of teaching us the most valuable life lessons. However, it is up to us if we want to understand them or not.

CHAPTER 11

Death North of the Greenough River

Warning for Aboriginal people: This story contains specifics relative to the death of a Yamatji person, north of Mullewa, WA.

I recently saw this picture of the Greenough River, 30-odd kilometres north of Mullewa. It was posted on one of the pastoral rainfall websites, and it got me thinking. This river was over a metre deep in April 2021. This is the sort of flowing water people try to cross in their vehicles and get washed away. Please don't try it, it's not worth it.

I remembered a very powerful experience I had close to this crossing around 1981. Margaret Hughes was with me. She was my jillaroo (house girl), and, as she had been stuck on the station at Curbur for a few months, I had suggested she come with me in the truck taking a load of goats to Robbs Jetty Abattoir south of Perth. The company and conversation in the slow old C 1800 Inter cab-over truck would be great. We were never expecting the experience we would have on this journey. No one would...

We were heading north after just fuelling up in Mullewa to complete the last stretch of gravel to Curbur, some 250 kilometres, after unloading the goats in Perth. In the early 80s, a slight stretch of bitumen ran north out of Mullewa on the 'North' road, as it was called, for a few kilometres.

GREENOUGH

It felt like it was there as a little teaser, if you like, to prepare you for the next 300 kilometres of rough gravel. And we were in an empty underpowered truck that often couldn't even get enough speed to get on top of the corrugations. You felt every little bump, rut and pothole after you left that short section of bitumen.

As we turned the corner some 30 kilometres north of there to cross the Greenough River near Tallerang Peak, we crossed over the dry low concrete crossing (in the photo above) and headed straight up the road, north, towards the peak. It was only then that we could see what seemed like two people, looking a bit like stick figures in the distance, possibly walking towards us.

'What's that?' Marg asked, pointing towards the distant sight through the large windscreen.

'It looks like a couple of people walking towards us,' I said, trying to make sense of what we were seeing. However, it didn't seem like a good omen of things to come.

They were still a distance away, but we could see they were not walking that well and stumbling a bit. As we got closer, I could see they were both Aboriginal blokes.

I stopped the truck to see what the story was. I saw they were both suffering from something and asked them what had happened.

'He motor he rolled ober, women he mulla,' one of them mumbled. 'He been accident.' As 'mulla' was the Yamatji word for dead, I said we'd better go and see if we could help.

'No, no, we not going back there,' they persisted loudly.

I wondered if anybody else was hurt and tried to get an understanding of what had happened. I guessed we were still about a kilometre away from the accident site. I asked again if anyone needed help, feeling anxious that things could be serious. As far as I could see, this was a critical situation.

As there was no room in the cab, I suggested they jump in the truck's stock crate so they could show me where the accident had occurred.

'No, we not goin' back dere,' they insisted, with fear in their eyes. Any dealings I had with Aboriginal people indicated to me they have a deep concern and suspicion of death or when people have died, so I felt I understood their concern and was prepared to leave them be for the time being away from the accident site. After some moments of contemplation, I suggested that they sit and wait for me on the roadside whilst I went and looked at the accident site for myself, and then I would come back and run them into Mullewa after assessing the situation.

So, they sat down on the side of the road on the small windrow while Marg and I travelled to the accident site, a little further north up the road.

The first thing I noticed was the circle work or burnouts from a vehicle in a gravel pit next to the road, with deep circles engraved into the floor of the pit. From personal experience, I could tell that the front wheel had been in a hard lock, while rear wheels had been spinning sideways with full revs, digging deep trenches into the ground surface.

There was a group of three or four people who seemed very much worse for wear sitting under some nearby trees, wailing and crying; over to my left, I could see an upturned vehicle that was partly in the bush.

After pulling the truck off the road proper to a point from which I would be able to safely risk turning it back around towards Mullewa, I went over to the upturned vehicle to see what had happened. It was clear by the amount of scrub crushed nearby and the damage to the outside panelling of the vehicle that the vehicle had rolled over a number of times. But my big shock came when I got closer to the overturned vehicle; I could see that a young girl, who had probably been thrown clear as it rolled, had unfortunately been crushed as the vehicle had come to rest. Surely, she had died immediately.

I could not move her from under the vehicle until it was lifted, and I could see that I was not going to get any help from the folk huddled nearby. They were all in mourning and involved in their own communications away from the accident site.

Marg was slightly behind me in surveying the accident scene, and I came back to her to suggest she did not need to see that. There was nothing else that could be done there.

In shock and disappointment, I surveyed the scene to confirm there were no other deceased persons around the burnout area. After a quick chat with the mourning passengers under the tree, and confirming that no one else was severely hurt and in need of a lift back to Mullewa Hospital, I returned to the truck to drive back to Mullewa and report the incident to the police. With a deceased party involved, I knew things were serious, as questions would be raised. I picked up my two roadside passengers from further down the road, and we all travelled back to the Mullewa police station.

The two passengers jumped off, and I quickly explained the situation to the police, based on the scant information I had. After getting details from the two men, the police overtook me in the police vehicle just before the crash site on our return trip north.

I stopped the truck nearby to assist. The two policemen, with all their strength, managed to hold the overturned vehicle off the deceased's body for just long enough for me to pull her body free.

I experienced that day what members of the police force undoubtedly experience regularly – the pain of seeing the loss of such a young life wasted and gone from this world. I'm sure she hadn't been driving, so she was probably not directly responsible for her own death. She had likely just been in the wrong place at the wrong time, though I was only guessing from what I saw. I was never made aware of the accident details later, nor did I feel I needed to know.

I have been very fortunate to have had many varied experiences in my life – some good, some not so good. That was one of the latter. There's nothing you can do but harbour the memories and try to learn from them, feel sympathy and understanding for those directly involved and try that bit harder to be a better person and more understanding from the experience.

Many family members in the Mullewa community and beyond will know of this accident, which occurred in approximately 1981, and its fallout. I'm hoping this revelation doesn't cause any more grief to them. I wished I could have done more, but we can only do what we can do in such circumstances.

CHAPTER 12

The Big Turkish Taxi Driver

When I left SJOG Hospital Subiaco after having my right shoulder reconstruction operation in 2021, a big, turbaned Turkish taxi driver picked me up out front. He started talking as soon as we set off. After several verbal attempts in broken English and much hand gesticulation, he finally managed to explain the point he was trying to make. And that was: 'How are you going to be able to wipe your backside with your arm in a sling?'

I replied, 'Well, do you write left-handed?'

'No, only a little bit,' he said.

I said, 'Well when you write left-handed, what does it look like?'

He said, 'He's very messy.'

I said, 'Well, you've got the picture then!'

The funniest thing for me was that it took some time for his imagination and understanding of English to kick in. But when it did, his big happy frame shook uncontrollably with laughter; not an easy thing to do, as there was barely space for the steering wheel to operate due to his large frame and belly. You might have called it a very considerable belly laugh.

I liked the fact that, with all the intercultural differences we had between us, he certainly got the joke! In fact, some kilometres down the freeway south, he still had some intermittent convulsing giggles. Even the car suspension seemed to be joining in. I was glad I helped to make his day, because the pain in my shoulder from the vehicle vibration was bloody awful, but at least one of us had a win.

So, my big Turkish friend, providing this book is published (and I'm assuming you will hear about this story somehow on the taxi grapevine in Perth, WA), should you decide to contact me through my website www.theflyingbushman.com, I would be happy to forward to you a personally signed copy of this book for your records.

Keep on enjoying our Aussie humour, my friend.

CHAPTER 13

The Old Cowboy and the Kid

Warning: This story may carry disturbing content for some readers.

'Lengthen ya stirrups, boy,' Tony yelled to me from across the cattle yards. I was about to get on this young horse in the big yard for the first time, and we both knew it would buck like shit.

Tony had been there and done it all plenty of times, and I took his advice, taking my stirrups down a couple of holes.

I was probably fourteen or sixteen. Tony Grant was perhaps thirty-eight or forty – in the prime of his life in terms of horse experience. No doubt he had forgotten more than I would ever learn. Just his presence and understanding of what was to happen settled me a little, and it was good to have him nearby in case of emergency.

I came home from work today and, grabbing a stubby from the fridge, reached up to fish around in the bowl on top of the refrigerator for Tony's stubby holder. As I did so, I remembered it was exactly four years ago, to the day, that he had passed away.

And I shuddered as I remembered the awful way it had happened. He lived not far from the Utakarra cemetery in Geraldton – 'In case he needed it', he jokingly used to say.

He must have been a mind reader.

I could hardly have called Tony a friend – he was nearly a generation older than me. Still, I distinctly remember the relationship we developed when I was a young kid and he came to Curbur to break horses in the old cattle yards near the homestead. As I sat quietly watching him from outside the round horse yard, hour after hour, I admired his immense strength when dealing with such powerful animals. A well-bred brumby colt with a lariat around its neck has the strength of a Mack truck when it comes to power on that rope. To be able to overpower that strength is nearly unbelievable.

I used to see Tony holding the lariat some 5 or 6 metres away from the horse's head and, with sheer strength, fight the animal to take control and pull the horse's head around to face him. Doing it myself years later, my hands would be wrecked by the rope burns and blisters, even with calfskin gloves. I wouldn't dare try and pull them towards me when they were facing away. I would wait until they were at 90 degrees to me, or better.

I remember as a young kid in the early 70s, when Tony and some of his mates came up to do some contract mustering for us with their giant, long horse floats, their paint horses, the big boots, the big hats and all the cowboy gear: spurs, ropes, American roping saddles and all the fruit. It impressed me as a bush kid as being the real deal. But those were just tools of the trade for them. They would regularly ride a buckjumper until he had no buck left or throw a big bullock in the flat and hold him down until a set of hobbles arrived. That was just what they did to get the job done; it wasn't any show.

Tony used to own and work on his own farm as a younger man, and was then on Yanget Farm in the Chapman Valley, just east of Geraldton, where the Grant boys had grown up on a sheep and cropping block. I remember him telling me that the young farmers around there used to take their girlfriends to dances in the 50s, carrying them to some of the beautiful old farmhouses in the district with a horse and trap (small cart on two large wheels used to transport people). I remember listening to him tell those stories and thinking how amazing it would have been to take a girl out on a trap in her finery, hair blowing in the breeze, jig jogging along to a party.

With his interest in horses, Tony became a rodeo rider; he had a 'bull string' (a group of bulls taken from rodeo to rodeo and used professionally) for the rodeo for a while and was a cray fisherman after he left the farm. Tony didn't stomach fools very lightly, but he loved the bush, the range land and raising livestock in that natural environment. He liked the bush people, which I think was what we had most in common in his later life. He liked the characters in the bush, their different ways of skinning the cat and their toughness, to survive in that environment with their families. That earned his admiration.

I used to drop in on Tony for a beer, as many did, in a few properties he had, including a small farm block he bought north of Mullewa. When I had the opportunity, we would have a few beers and talk stations, the news of pastoral people we knew, and who was doing what. In later years, he used to do a bit of caretaking at stations around the Meekatharra area when managers needed a break.

He would tell me about their excellent herds of cattle, and the better pastoral herds, and how the numbers of dingoes were increasing and devastating the animals in the pastoral country.

I also think my father appreciated Tony. Dad wasn't the sort who handed out accolades very readily or got too excited about people's abilities, so if he did, you were inclined to feel that they were pretty good. Given that Dad gave him the nod, he was probably worth considering.

Knowing he always loved to work with livestock in the bush, I asked Tony to be a steer ride judge at the BBB (Ballythunna Bush Bash), event in 1994 which we were able to organise on Ballythunna Station in the Murchison District of WA, a property that I owned for twelve years. We had the cattle for the rodeo, the magnificent wild brumbies, and, once we built the infrastructure, we had an annual event to raise valuable funds for the RFDS and to recognise all these people's efforts and many more unsung heroes in the health industry. This was a few years after my accident in the Pilbara, where they not only removed me from the accident site (as per back cover) but flew me to urgent medical access in Perth and surgery for spinal fusion. Many thousands of dollars were raised during the two years of the

BBB and it was all donated to the RFDS in recognition for the help they gave me, no doubt saving my life.

Horse breaking in the 70s was more about breaking a horse's spirit, I guess, and physically showing them that man had superior strength and (supposedly) intelligence. As a result, they had to succumb. I appreciate that training horses now has become very different. Today, you spend more time working with the way the horse thinks and using that to your advantage.

I knew that Tony could be a hard man at times. I gather he held his own in the rodeo circuit, and they were tough men in tough times, but I was too young to witness any of that. I watched him work breaking in some fifteen or twenty horses that he had selected, as he had done some per-head contract deal with Dad to break in camp horses for the mustering camp, while I observed him closely.

So, as I was his understudy, the time came for me to climb onto this young horse. We both felt there was every chance it was going to buck its skin off.

After he hollered at me to lengthen my stirrups from across the cattle yards, he said, 'I reckon that bloke is going to have a go.'

As soon as I got on, the horse squealed and, with half a dozen bucks, pitched me to the shit... I had thought I had him for a bit, but not on your life.

Tony could read the situation and caught the horse over in the corner of the wooden yard while I found my feet and got my breath back. He said he would hold him while I got on and hold his head for a bit. We had a strong rope halter under the bridle to have better control of the youngsters, and he clipped on the lead rope as I proceeded to take up the reins.

The colt bucked again, all right, but as he was trying to get into full flight, Tony just kept hold of that rope and his head. Since those pig roots were the best the horse could do, even I could hold on.

Eventually, after a few loops of the yard, most of the sting and buck were taken out of him. Tony unclipped the halter rope and stepped into the centre of the yard.

'Step into im, boy', he said, and after a few steady laps, I managed to work the tickle out of him, and he settled down well.

'He rode him alright,' Tony said that night over a beer before dinner to Dad. 'No worries about that.' I think I pushed my chest out and had an extra mouthful of beer with an embarrassed grin, secretly knowing that Tony holding the horse's head initially took most of the buck out.

That was the thought I had when I pulled up in front of Tony's place on 19 June 2015. I had called him a few weeks before to say that I would come up and see him in Geraldton so I could catch up with him and see my daughter and my grandchildren. I was coming from Capel, so I had a fair drive, some eight or nine hours. I wanted to give him a bit of a surprise since he had checked himself out of the hospital, much to the doctor's frustration, with a cancer diagnosis which wasn't very good.

On my arrival, I noted that his old blue heeler, usually full of bark at visitors, was very quiet. His greeting was half-hearted, and I thought, surely, he didn't remember me. It'd been a while since I'd visited. Or maybe he was off-colour.

I yelled out to Tony while still outside the front gate, not entirely trusting the old dog, but certainly feeling his demeanour was off. I yelled a few more times to no avail and figured Tony must be out the back.

I opened the gate gingerly and moved towards the side entrance we used to take, but the dog was not interested in me. He had other things on his mind, I could see.

The door into the rear kitchen was open and, still calling out, I entered, continued into the lounge room, and inspected the bedroom to confirm that he wasn't in bed unwell.

His bed was unmade, and the bedside table light was on. It seemed unusual he would have left that on during the day. Walking out the back, I thought he might be doing some gardening, but there was no answer to my calls. I also came past his Toyota 4WD tray top, where he used to take the dog on a beach walk, parked in the garage, so unless friends had picked him up, I figured he must be here somewhere. Perhaps over at the neighbours.

I was starting to run out of ideas when I looked to the back shed at the rear corner of his block. Perhaps he was working out there and had not heard my calls, although his hearing was pretty good for his age.

As I started to walk towards the door of the back shed, the old blue dog was with me, right at my knee, practically pushing his weight against me in his keenness to get that shed door open. Something was going on, but I didn't know what.

When I opened the door, the dog forced past me like a freight train. I don't think I could have stopped him even with a lead; he was hell-bent on getting into that shed. Nothing was going to stop him, and, soon enough, I understood why.

Even as I was wrestling with the dog's reaction in my mind and wondering what was causing it, my body was in motion, and I had walked into the shed. I didn't entirely know what to expect or what I would see and was only half concentrating on what was in front of me anyway. But I was utterly stunned into the present at what was my first experience of this type of suicide, which is brutal for any witness to experience. Like so many proud old bush blokes, Tony had decided that his time was up from the cancer, and he wasn't returning to hospital again.

The problem was that, unfortunately, I was the one who found him. He'd chosen to use his own old unlicensed shotgun to end his life in his back shed. Not an experience you would wish on your worst enemy.

Tony, during his life, didn't ever do things in halves; he continued that in his death, and he left no chance of anybody thinking he had been half-hearted with his intentions.

Like me, the old dog was trying to come to terms with what was in front of us. The shock was all we could confirm at present, but his sniffing around got me grabbing him and, despite his reluctance, putting him outside and shutting the door. That's no doubt what Tony had done with him not long before; shut him out of the shed. He would have heard the terrible gun noise and then the silence, which he would have guessed was not good. No wonder he wanted to get through the door, in an attempt to make sense of it all.

After quickly surveying the awful scene to ascertain the situation, I realised I needed to call the police, and went outside to do that. But as I made my way to get my mobile phone, Tony's nurse from the hospital pulled up, out on her regular visits to him. Appraising her of the situation, I suggested that she ring the police, as she was his nurse and carer and could make things as smooth as possible under the circumstances.

The police arrived and asked me standard questions and whether they could take a statement. I said I would need to leave it for a bit, as I was still suffering from shock and disbelief. Another mate of Tony's arrived, and I knew he would give the next of kin and family the bad news.

Tony's old dog came and sat alongside me outside, and we cried together. I hugged him one last time for Tony, and he barely moved. He, like me, needed something tangible that we could touch and feel and hold after that experience, and try to put life and death into context and try to come to terms with it. Something to touch and feel that was alive and could give and take affection, be responsive, because we had just been in a terrible place.

Tony's mate organised for somebody to care of his old dog, somebody he knew well.

My brother Keros went to Tony's funeral on our behalf to represent the family, as he had visited him in hospital just prior to Tony's checking himself out and was one of the last family members to see him alive, I guess.

I couldn't make it back to Geraldton for the funeral, as I was now back down in Capel in the southwest for work.

But every time I have a drink out of that stubby holder, Tony, I realise that horse never had a chance in hell of getting a buck in with you holding his head – not a chance in hell.

Keep ridin' em, old mate.

RIP, Tony Grant.

CHAPTER 14

Predicting Two Generations' Future from a Teacup

From early childhood, my mum, whom I was very close to, regularly told us a story about how her Aunty Dorie read her future one day from a teacup. Mum was very spiritual; she believed in the afterlife and all that the universe had to offer. More to the point, Dorie was dead right, as everything she foretold came true over the next sixty-odd years.

It all unfolded very much like this:

As mentioned in some of my previous writings, Mum was a professional singer in Adelaide in South Australia when she was a young woman in her early twenties, before she was married. After a rehearsal one day, whilst practising with her pianist, Aunty Dorie, as Mum always called her, they stopped for a break and, as was often the case in those times, had the obligatory cup of tea. Of course, we're talking a proper cup of tea here, with the teapot and leaves and no filter, none of this dip dip stuff!

Now Mum was aware that Aunty Dorie read teacups and was conscious she had given it a break for some time because she was foreseeing things that weren't that pleasant for people to hear, such as the pending death of family members. She gave up the readings for some considerable time.

But on this occasion, perhaps with newfound vigour from the positive rehearsal they had just had, she asked Mum to give her a look at her teacup as they finished their cuppa. Mum passed her empty teacup over, probably not realising that Dorie was reading so far into her future and that of her children. I'm guessing Dorie would have given the cup the upside-down treatment, as Mum used to do years later.

You must realise that my mother and father had not yet met at this stage.

Aunt Dorie's reading was as follows:

She said, 'I can see a tall man here that you will marry. He seems to be in uniform with a little peak hat. Your husband will have a business partner [she described him and his personality], you will be with him for a long time, and you and your family will move far away.' She described the front of a homestead or large house, with its pitched green roof and other details. 'You will have four children, two boys and two girls. The girls are older and will move away from home, and the two boys will run the property, but the eldest will move away and do his thing, while the younger boy will eventually take over the property. I see older women with long dresses and big hats. It's very hot and dry, water is limited, and there's little shade.'

There were a few other incidental details, but that is the bulk of the story Dorie saw in Mum's teacup. Mum repeated that story many times over the years to us when we were children, from our early childhood until the time I was probably twenty-odd. The story was always the same, as it no doubt impacted Mum at the time and, as it played out in real life, would have probably gained an even greater impression on her.

But let's dissect this story a bit more so you can grasp what Dorie perceived and saw.

A year or two later, Mum met and married Dad. He was six-foot-something and was in his Air Force uniform, with a peaked cap. Dad did have a business partner for many years, Wynn Brandon Brown, who came from Perth and then moved to Adelaide before his death. They were business partners for about twenty-five years, and Dorie's description of him was spot on. Wynn was a goer and wheeler-dealer but mostly a prick, although her definition was much kinder.

Outback WA is a long way from Glenelg in South Australia. Dad came over originally to Darkan to manage and clear a beautiful place known as Dunleath Farm and then, after eighteen months or so, went to work Curbur Station in the early 50s. Mum had the two girls in Adelaide before coming to WA, and then we boys were born in Geraldton, WA, and returned to Curbur Station, which we had purchased from Wynn's family in the 70s.

Of course, the ladies in the long dresses and hats were the Aboriginal or Yamatji women, who in those times did wear the long old dresses and often old cowboy hats or similar to keep the heat and sun off in the desert where we cut our teeth. And yes, the girls did grow up and get married and moved to Perth and elsewhere. My younger brother Keros and I took over and ran the station together for a while after he left school.

But here's the interesting issue: Does this mean you cannot change the future? Are we all just puppets on a string? Or do we make our own life happen, or is it a combination of both?

That's the unanswered question, I guess.

Anyway, the point is that, as I mature (well, hopefully regardless) and use my own life experiences of having been to the other side and back again twice, I feel confident the universe has a big part to play. Especially when we are younger and full of energy and vitality, we swim and thrash against the tide and push and shove – often to no avail – and what happens, happens anyway. Maybe it would have been more sensible to just rest in the shallows and wait for the tide to change: What do you reckon?

I'm sure my brother Keros won't mind me saying this, because he also heard the same story for many years. However, it bothered me a bit when Mum retold this story, and Keros was still down in Perth at boarding school while I was working my butt off on the station to keep things together. I couldn't help but wonder, how the hell does he end up with the station after I busted my butt there? When Dorie made this call, it was potentially way into the future, but she had been spot on so far.

Well, that was precisely what the universe had decided, and, in a way, hers was a more relevant picture than Keros or I could have ever imagined it to be.

I moved on and developed a state-wide helicopter mustering business, which I wrote about in an earlier book, *The Flying Bushman*, and Keros and his wife Simone proceeded to take over and run Curbur Station.

And that was precisely where Dorie's reading in the late 40s ceased! She had predicted from the tea leaves in one cup some sixty years ago pretty much exactly what would happen. Pity they didn't have lotto back then.

I wonder what I looked like in Aunty Dorie's tea leaves when she said, 'He'll move away and go and do his thing.' I guess there weren't many helicopter musterers around in those days.

And, as they say in the classics, 'It was all written, not in the stars, but in the tea leaves.'

CHAPTER 15

'Your Mother Saved My Mother's Life.'

I was way out northwest of Meekatharra in around 1998/99 when a young Aboriginal woman, Priscilla Simpson, leaned across my interview desk and took my hand in hers at the Yalga Jinna (Jinna means 'foot' and Yalga means 'flat' in the Yamatji language) Aboriginal Community 130 kilometres NW of Meekatharra in WA and said, 'Your mother saved my mother's life.'

I was out there completing a state-wide survey for Princess Margaret Hospital on Aboriginal Children's Health on my consultancy. Although the comment shocked me, I realised what she meant after I gathered my thoughts.

Her mother, Margaret Simpson, was born and bred on Curbur Station in the Murchison, as was I. This was our home country. She was an Aboriginal woman, and her family and parents used to work for us on the station for many years from the mid-50s to the late 70s.

Margaret was a Yamatji girl, one of Eenie and Johnny Simpson's children, and she was playing down near the rubbish dump as children did then, where the old vehicles and carts and things were ideal for children to play on. She was with her younger brother Lawrence who was digging up old cans and bottles from the old historical dump.

Unbeknownst to anyone, an old dingo trapper had thrown a bottle of his old, out-of-date, pink strychnine tablets into the dump years before, not expecting anybody to go excavating. A lot of station hands and doggers had strychnine tablets in their kit bag in those days and used to lay dog baits out on the property all the time. Unfortunately, it seems Lawrence found a partially buried bottle and managed to open it in search of the attractive pink tablets.

We can only guess, but he probably consumed some before his sister Margaret consumed some, and it was then that they or somebody raised the alarm. They were rushed across a few hundred metres to the homestead where my mum, June Keynes, regularly cared for their ailments. However, on this occasion, Lawrence seemed to have either consumed more strychnine or taken it earlier than Margaret. Although urgently trying to make the children vomit with salty water or whatever, Mum managed to make Margaret sick and vomit up the poison but, despite desperate attempts, could not save Lawrence as he became completely unresponsive.

I don't think I was there to witness the awful experience because I would have also been very young, but I do remember the impact the experience had on Mum, and she spoke of it tearfully for some time.

But it was so fortunate she was able to save at least one of the two children and for me to have that experience with Priscilla. It meant Margaret had children, and I know for a fact there are numerous great grandchildren.

Mum, if she were alive, would have been so thrilled to see how her bush nursing act gave so many children a chance at a healthy, happy life.

CHAPTER 16

To the Other Side and Back Again x 2

Many people ask me about my near-death experiences (NDEs) of going to the other side and living to tell the tale, as it were. There was always a lot of interest in this topic during radio and magazine interviews, so, I thought I would be more specific in this book about exactly what happened and what I experienced in having the opportunity to visit that amazing place.

In my previous books, I touched on those experiences: Once as a child, and another time when, as a twenty-four-year-old in 1984, I had a helicopter accident, crashing around fifty metres onto a rock floor. The accident was so bad that specialists didn't know if I would ever be able to walk again or have children. So undoubtedly, it was severe enough to fit into the NDE category and to take me to the other side and back again.

There are some things that stories of NDEs often have in common, such as floating above the bed in surgeries, where doctors and nursing staff are around the body the storyteller once inhabited. There are generally spirits there who can speak directly into one's head, and the dying person completely understands what they are saying. And when they get to the other side, whoever is talking to them seems to be doing it without a voice. It's also often said from these experiences that, when you advance into that other realm, you are usually guided on your journey by your deceased relatives.

Some of those experiences happened to me, but I'm sure everybody's experiences differ and are hence so interesting. But how can you supposedly go to the same place and have a different adventure?

I can only speak for myself: I believe entirely in near-death experiences and all the different examples that individuals experience, because they are the only ones who can describe what they experienced.

Of course, there will be the odd person who embellishes that experience, for reasons best known to themselves. Still, I feel for those who have legitimately walked the walk and feel that the spiritual power of particular experience impacts one to such an extent that there is no need for embellishment or fabrication.

I felt impacted by that beautiful consciousness and began to see the world more compassionately and with more understanding and respect. I'm more reflective, maybe, but the compelling impact of the experience is what always remains with me.

I heard a lady recently saying how the pain of the death of her son in a car accident, as terrible as it was, reduced over some significant time, as the memory and details of the event lessen somewhat, but that her NDE, at the time of the accident, was as strong today as it had been twenty years earlier. I agree with her – except mine was forty-odd years ago.

However, it's essential, I think, to understand that NDEs are extremely difficult to describe or explain. What you are feeling and seeing is not always easy to put into words and relate to people, partly because what was seen and felt is entirely out of this world. Therefore, I will use a few metaphors and similes to describe what transpired for me, so that you can grasp a better understanding of the experience.

Many people don't wish to talk about these experiences at all, often perhaps because of the impact it has had on them, but also because of audiences' mixed or adverse reactions. Why would you bother trying to convince someone who's sceptical of such an experience?

In my case, I was the only one present at my NDE. There were no doctors or nursing staff to confirm my stories in a public place like a hospital. It was just me, in the middle of nowhere, having that very personal, powerful experience with what I am sure was the afterlife.

Nearly forty years later, I am still amazed how much of what I experienced had similarities to what others have said about their experiences. But although similar, in another way, they are all slightly different, as people from different cultures, backgrounds and ages relay their personal journeys. It's what they saw and what they interpreted through their eyes, soul, spirit or whatever.

Scientists generally agree that there are a few criteria that define an NDE and are usually involved when the individual's life is in jeopardy. They are as follows:

1. An out-of-body experience – where you float above your body, for example, on an operating table
2. Pleasant and loving feelings in that peaceful, caring environment
3. Seeing the light in a tunnel, a bright light, deceased relatives or something similar
4. A review of your life
5. A conscious return to the body

It seems you don't have to have all these experiences, but some or most are relevant to whether you have had an NDE. These points were developed by scientists studying this phenomenon for fifty years or more.

When it comes to what I experienced as a very young child, with more consideration over time, I don't think now that it could be defined as an NDE. This is because I didn't feel entirely like I went to the other side. And although the experience was authentic and frightening, I don't believe I crossed over. I was saved from going there by another powerful force, I'm sure. And that force, supernatural being or presence enabled me to stay on earth, in the paddock. But that presence did overcome me and helped to settle and relax me considerably and remove some of the stress and anxiety from my experience at the time after I had slumped into a space of complete and utter desperation. Maybe someone up above considered the complete NDE would be too much for a young kid to handle and decided it was best for me to keep my feet on the ground.

With my later experience of the afterlife as a young man, I'm sure a child would have absolutely nothing to fear and would be completely bathed in the love, power and beauty of the experience. They might explain it more straightforwardly, but it would still be as vivid and impactful.

If you do happen to explain the journey to people, it's because you want to do it justice and capture the gravity and intensity of the experience. But for me personally, other than writing about it here, I do require a very respectful listening audience to share with, which is extremely difficult to find collectively. It's not something you would or could discuss over the bar with the boys at the Sunday session, though, if somebody can do that, hats off to them.

I am not saying for one moment that I am an expert on this topic. I just experienced what I experienced at the time. And although it happened ages ago, I can remember it as if it was yesterday. I don't even understand why, in 2022, we are even bothering to contest the validity of these experiences. So many people have had them, and now, on the Internet, they are all there for people to read.

Let's say, for example, you see a vision of dear Auntie Jane, who you haven't seen for years, as clear as a bell in front of you, from entirely out of the left field, when you are shopping or sitting on the toilet. Then you get a call from a hospital on the other side of the world to tell you Auntie Jane has passed away, pretty much at the exact time you had your vision. What's so surprising about that? She just wanted to say goodbye!

Maybe we don't know all the wheres, hows, and whys, but is that so important? Perhaps for the scientific mind, it is. I guess a lot of what we believe and think has much to do with how we are brought up, our parents' beliefs, what we are conditioned to think and our childhood surroundings. If our mum or dad boohooed these topics around the kitchen table, we are probably less prepared to consider them now.

However, I'm sure there is more consciousness today, and people are generally more accepting of the 'stuff that happens on the other side'.

What do the Americans say so well? 'Get on with it already.' So, let's dig a bit deeper into this subject and see if we can get an improved understanding together.

* * *

As I said previously, an extensive and powerful NDE happened to me when I was twenty-four years old and crashed my helicopter into a gorge in the Pilbara. The level of severity of that horrific accident was certainly enough to take me to the other side. The phenomenon of NDE usually only occurs when an individual's life is in jeopardy, as mine certainly was. You don't go there for a toothache.

When the engine failure occurred, I was flying low and slow, or hover taxiing, with the helicopter at around 150 metres AGL (above ground level), checking the gorge sides for cattle escape routes. I had just relayed my thoughts of using the gorge for an overnight camp to Rodney Pulford, the pilot in the other helicopter, and my ground crew, via UHF radio, so that we were all on the same page. I asked if somebody else had a better idea.

I wish they had, as I might have been flying over better terrain on which to land my faulting machine, rather than where I ended up!

When the aircraft coughed and lost power, I began to drop at an alarming rate. I screamed expletives over the radio, I'm sure – probably something like, 'F–, I'm going down!' Rodney had a pretty good idea of my rough location, as we had been discussing the plan of where to camp the cattle overnight just moments before. And no doubt the 'going down' comment and the tone of my voice would have left little room for either Rodney or the ground crew to misunderstand the gravity of the call.

As frightening as the rate of unexpected descent was, I realised instantly that there were a limited number of places where I could land. It was the worst possible place ever to complete a successful autorotation or engine failure procedure and land without considerable damage to myself and the aircraft, as there were upturned outcrops of rocks a few metres high everywhere I looked.

As I was falling like a stone, I realised I would probably not make the top of the hills overlooking the gorge and would be forced to crash into the vertical side of the canyon or a sheer cliff, guaranteeing disaster. At the same time, I couldn't choose my ideal landing pad.

Without any more options, I manoeuvred the sluggish machine as best I could towards the bottom of the gorge, hoping for a better alternative. But as split seconds rushed by, control of the aircraft dissipated, since I had no drive motor. I realised I was entirely out of options. There were no split seconds left to get it down the best I could. I remember seeing this large outcrop of rock hurtling towards me and realising that this would be serious. All control had been taken away from me; it was like I was riding a horse without a bridle or reins. Nothing I could do would make any difference. Despite my flying hours and pilot experience, I was defenceless.

Out of power and all out of aces.

Now, this is about when other powerful forces started to kick in.

In that moment, with hopelessness reluctantly dawning on me, I was consumed and overcome by a powerful calming feeling. It was a bit like being conscious in two adjacent worlds. One, inside my body, where these massive outcrops of rocks were hurtling towards me and knowing there was a huge smash coming and serious pending harm to myself and the helicopter. I was almost wincing at the oncoming impact. The other, outside my body, was watching as if somebody or something had removed me from the repercussions of what was about to unfold, and that image was sitting over and outside the bloke in the helicopter.

I must have performed an engine failure procedure automatically, the best I could, because one rotor was practically the only item undamaged in the accident. That's something you learn in your training to get the aircraft on the ground intact. It was later cut off and used in an old mate's den as a wall ornament. A bit of a talking point over a Scotch with the mates, no doubt. So, that confirmed that I must have gone through the procedure, even in that state, but I don't remember any of it.

A consuming peaceful sensation enveloped me despite seeing the impending accident. Part of me was bracing for the crunch landing, but the other part seemed to be very much at ease, as if the real world for me wasn't going to be affected by the outcome of the crash. It was as if I wasn't present for that colossal crunch but was on a journey somewhere else.

I felt like I had smoked several marijuana cigarettes just before the incident; you know it will hurt, but you're not stressed about it because you know you won't be able to feel it. Maybe it's nature's way of taking away the severe pain of a situation before it happens.

I can barely remember the actual impact. There's just the slightest memory of the ripping effect of the crash landing on my body and the noise of the crunching machine. It was like watching a crash on TV: I could see the accident, the impact and the crumpling of the body, but I didn't feel it, as if I weren't there.

I would not be able to guess what my heart and brain functions were doing at that stage, and it probably would have been the last thing on my mind. Indeed, after that, everything went blank. I presume I fainted or similar. I awoke immersed in this serene, peaceful environment. I don't know how I got there; it seemed I had just arrived. You've no doubt heard the saying, 'walking on air'. It did feel like there was some of that in my experience.

No deceased relative was leading the way, and I didn't run into any of them at any stage of the proceedings.

You know how it feels when you go into a place and the atmosphere completely takes your breath away? It may be good or toxic (in the earthly case), but whatever it is, it engulfs you. It completely consumes and overwhelms you and your emotional state. It's like walking through a thick fog or smoke – you can't possibly miss it.

That's what this was like with the overwhelming feeling of love, kindness, peace and respect.

The only words I need are that this was a beautiful place. I could reach for other adjectives, but that's all you need. I don't think there would be a human on the planet who would not have enjoyed my experience. That's probably why it has such a positive and long-lasting effect on anybody who has experienced it. Take the best and most enjoyable experiences you have had in your life to date, add those experiences together, and multiply the result by five or even ten. You will then understand the positive impact these NDEs can have on people. Add to that combined experience a touch of out-of-this-world universal power and wisdom, and you're starting to get into the 'space' I'm talking about.

And, as it says in the big book, it was 'passing all understanding'. I'm not religious. But that is the best definition I could give that place, as it was passing all understanding, for us humans anyway. You couldn't imagine such a beautiful, loving, peaceful place. My spirit, my soul or whatever state I was in at that time would have been happy and content to spend the rest of my days there. There was nothing to fear and a complete loss of the fear of death.

Looking back, I think this was the experience of the journey which was so enduring and influential. This was the impression that was left on my soul or spirit; spending time in that space and having its influence rub off on me.

I have never experienced such a place before or after, but it was a complete feeling of peace and relief. Believe me, if or when you get to this incredible place, you will know it without any doubt. There's no guessing, you will just know.

I imagined this to be heaven or the space before heaven, because, in what seemed like a waiting room, I was constantly drawn to a compelling light. It was like somebody having the most powerful headlamp on a dark night, shining so brightly I could barely see precisely what or who it was radiating from or see my footing. Still, I knew I didn't have to worry.

It was blinding in one way and not blinding in another, because although I could not see the being it was radiating from, it was a 'cold' light that didn't burn into my eyes or hurt, but that radiated power and supremacy. However, and I think this was unique to my experience, there was one request that I did have: this 'mantra'. I will never know from where it came, from within me or from an outside source, but I was constantly saying it, like a drunken sailor.

As soon as I realised I was heading towards that light, I began saying, 'I don't want to go, I've got too much to do, I don't want to go...'

I distinctly remember saying this repeatedly to whoever I was relating to in that beautiful environment. I didn't even know exactly where it was that I didn't want to go, but I knew distinctly I didn't want to go 'there'.

Now, this bit needs explaining:

I said before it was as if the standard communication between all parties in this place came without words or speaking. Somehow you just got the message into your head. All the communication in this place seemed to work in the same way; it was as if you knew what they were saying or feeling, and in the same way, they knew what you were saying and feeling – all without a word being spoken, or any noise or gesticulations made.

There were none of those noisy, raucous, disrespectful, voice-raising communications used at confrontational times by us humans – nothing but that peaceful, respectful, beautiful, utterly succinct communication. It was the most efficient communication you could imagine. It helped set the tone for this quiet and peaceful environment.

I saw the vague outline of a very regal, serene, and spiritual being in white robes, with 'his' (I definitely felt it was Jesus) robes over his head, and he was certainly male, with just the centre of his face showing. Because of the authoritative light radiating from this heavenly being, it wasn't easy to make out anything clearly.

I say regal because power and authority were radiating out from that presence like the light was, as well as love and good intentions. The draw of the light was overwhelming and constant. There was no way you could detach yourself from it. It was like being carried forward on a conveyor belt. It didn't matter if you tried to stop or delay your progress; you were going to end up at the feet of Jesus when your time came.

I was under the control of that being with the powerful light, and although the force was not frightening, I knew, without a doubt, that I couldn't overpower it or get away from it. But I also felt that there was nothing I could do wrong in this place; acceptance was all around, and there seemed to be other people in robes or white clothing who were seated and just murmuring in a muffled whisper that I couldn't understand. It was like court members in an old all-white laundry powder commercial, because everything was white and clean, although I did feel I may have been shuffling along on a dirt floor, or perhaps more like river sand. It was clean, smooth and not dusty.

As I was drawn along like a box on this slow but steady conveyor belt – the only way I can explain that 'powerful draw' towards the majestic being emitting the light – I knew there was only one place I could end up. The power was immense; there was nothing similar in our world or experience to compare to it. It was an external force of tremendous strength.

I was never concerned about what was happening; it was altogether out of my control. And I didn't feel threatened about what would happen when I eventually approached the light.

I was totally at ease and at peace. As they say, I had 'not a worry in the world', perhaps because I wasn't in this world? The other souls around me were all loving and respectful of my journey, but we didn't converse or communicate. It was as if they were on their journey, and I was on mine.

There were no instructions to be quiet, to sit down or anything as elementary as that; everybody seemed to be performing their duties as required, by whom I didn't know. But it didn't require communication at that point between anybody other than those members of the 'court', who were mumbling quietly amongst themselves. It was like we all knew what we were meant to be doing and were just getting on and doing it.

But I could still feel the love and care in the room – it was like the reverse of a public bus stop where everybody was selfishly thinking about their little world only.

As John Paul Young sang – 'Love is in the air... in every sight and every sound...'

When I eventually got close to the powerful light on my transparent conveyor, after neither a fast nor a slow journey, I had the definite impression that this was the place of no return.

There were only two options. You were either in or out. There was no angst about this decision by anyone or any being – both alternatives seemed to be a good choice. They weren't right or wrong – they just were.

Like a sheep at the pivotal point of the drafting gate, I knew I was never returning to where I had come from once that decision was made. There was no right of reply if I made that decision or had it made for me.

I was continuing with my mantra, of course. 'I don't want to go, I've got too much to do, I don't want to go.'

Remember, I wasn't saying that out loud – I was relaying it, but I knew that everyone had absolutely got the message; there was no doubt.

By now, I felt I may have driven the establishment mad. God, what do we do with this fellow? I didn't want to leave any stone unturned, however.

The love and peace from this regal being increased the closer I got, but at no time did I feel threatened or frightened. With all the positive feelings and vibrations radiating around the environment, there was no chance of that ever being a consideration. It was just a fact of coming into proximity to a regal being. It felt like Jesus, or what I assumed to be him, but I think it was how he and his imposing presence seemed to oversee everything that was happening.

You could say there was a moment when I was in his consideration, nearly able to touch him. In his brilliant white robes, the light was a little less intense, or perhaps I had gotten used to it. I say 'he' because it seemed like this was what Jesus looked like in his male form. Or maybe it was just my imagination at work, and I saw what I thought I should see, but I don't think so.

The environment was too majestic and too powerful for any guesswork.

I was in the closest proximity to what I assumed was my creator; that was certainly the vibe that I was picking up from this authoritative being. I have no recollection of direct communication or face-to-face contact, which I must admit seems a bit of an anti-climax, even for myself, after that journey. I don't remember a smile, nod or any acknowledgement.

Thinking back, I believe that would have been the cherry on the cake, but it wasn't to be. It felt like I didn't get any acknowledgement at all from JC. However, because I was still repeating my mantra of 'I don't want to go, I have too much to do', perhaps he had already made his mind up about me. Maybe I had already selected my draft gate, and hence communication wasn't needed.

He may have been otherwise occupied; that's how it seemed anyway.

I would have liked to have met him, but I guess there will be another time for that, and maybe we will have time for a face-to-face because I'm sure my mantra will be different. Should I have one next time?

Maybe he instinctively knew I was 'hell-bent' on returning to Earth. (Not a bad pun under the circumstances, ah?)

* * *

Today, I have three children and four grandchildren; I've kicked a few little goals and hopefully, there are many more to go; and I've done some writing, so I guess if I had gone left instead of right, or He made decisions differently, I wouldn't have been able to complete those joys I needed to achieve.

After being in that beautiful, peaceful environment, returning to reality on earth was a shock.

The next thing I knew, I was in that awful, painful, shocking environment when I came to in what was left of the wreck of my helicopter, and Rodney was trying to gently extract me from it.

I felt like saying to him, 'You'll never guess where I've been', but I figured it would be too hard to explain. But shooting pain was taking away my sense of humour and bringing me back to earth. And I was trying to grasp exactly what had happened. Although I felt I knew most of it, some details were a bit scant at that stage.

It was like waking from a wonderful dream, which has taken your heart to the heavens, and then waking up hanging off the edge of a cliff on a thin strap. It's a bit surreal.

Rodney had come and found me on my rocky outcrop. It was initially difficult for him to find a place he could physically climb down the sheer face to reach me, as around the crash site there was nowhere he could consider landing. Meanwhile, I had been around the universe and back.

The first words I heard from him now that I was back on Earth were, 'Come on, mate, let's get you out of there.' He was concerned about the aviation fuel leaking from the mangled aircraft, which could explode into fire.

I feel that there was every likelihood that somebody might have been looking after me there as well, because the fuel leaking over a smashed and exposed hot engine could have been a further disaster. Meeting JC twice in the one day would have been fantastic luck, but I think He planned it accordingly.

I was feeling very disoriented at this stage, not only about where I found myself, but about the recent astral journey I had just been on. I could not at once appreciate the gravity of the crash or the damage that had occurred.

The helicopter, which had been around three metres high to the rotor head, had been compacted to just over a metre high, with me still inside it. So, I'm sure Rodney did a great job carefully removing me from what was left of the wreck, or perhaps he just wanted to confirm I could still write his next wage cheque, ha!

When he finally removed me and partially laid me out on the rock slab, my legs had no feeling, and the pain was now really starting to set in.

I was very fortunate in another critical way, as it turned out.

My previous helicopter instructor, John Evans, had taken a job flying larger helicopters offshore from Karratha after giving me my helicopter licence at Jandakot in WA. And I was very grateful for his considerable experience as an instructor, having come from the Police Air Wing in London, where he had worked for several years.

Before the accident, he had invited my ex-wife and me to Karratha for the Christmas holidays. He had kindly introduced me to all his fellow Bristow Helicopter pilots over the break. As soon as Bristow pilots heard the RFDS emergency radio call and remembered me, a fellow helicopter pilot from the earlier Christmas holidays, they kindly jumped to and, after locating a doctor, flew down ASAP, taking about an hour to reach us. At first light, they were there in a Bell 206 helicopter to winch me out of the gorge once they managed to locate me, which wasn't easy and required communication from Rodney's chopper. But I'm sure they would do similar for any emergency when the RFDS called them.

That was a lifesaving decision because, although nobody knew the extent of my injuries at the time, once the doctor examined me on site, he knew there were significant spinal injuries. And a trip via 4WD on that rough terrain to get medical assistance would have killed me.

In short, after sixteen hours in the gorge, with just one Panadol for pain, they got me to medical treatment and onto the RFDS jet to Perth before a substantial L4 spinal fusion operation by six neurosurgeons, including Dr George Bedbrook and others. After many months of physiotherapy and with a neck-to-knee plaster, I could walk again, slowly.

I would rather not spoil this NDE chapter by recounting that difficult, challenging time in Ward 11, the spinal unit at Shenton Park, but it's all there in my second book, *The Flying Bushman: An Australian Story of Life above the Land*, should you wish to read about it. But certainly, from me, love and best wishes to all those tireless nurses and doctors who are so committed to their cause: We're all so grateful for what you do daily.

The doctor and crew carefully removing me on a stretcher out of the gorge.

Look at what's left of the aircraft they removed me from. (The front of the helicopter is closest to you.) Then look at the sheer cliffs in the background I was trying to avoid crash landing into, which are around 50 metres high.

I still wonder today what the universe had in mind for me with that crash. Was I pushing too hard, too fast or trying too hard? Was I spreading the safety tape a bit thin, or was that just a meeting I was meant to have? And if so, what am I meant to do with that experience? Or was there a message I completely missed?

Perhaps I need to consult Auntie Dorie's teacup to see what the future holds.

Perhaps part of it was to write a book and inform people of what happened on my journey.

I reckon I'll get the message from above at some time to explain the background.

Next time I meet JC, I'll have to pose the question, and I hope we have time for a chat.

If you have a near-death experience, I hope you get the correct gate for your future and end up where you need to be. He doesn't miss many on the draft.

In conclusion, I very much hope that you have enjoyed these bush stories.

– The Flying Bushman.com

www.ingramcontent.com/pod-product-compliance
Ingram Content Group UK Ltd.
Pitfield, Milton Keynes, MK11 3LW, UK
UKHW062313290726
14090UKWH00018B/1038

9 780645 669701